CENTRAL PLACE THEORY

LESLIE J. KING
McMaster University

SAGE PUBLICATIONS
The Publishers of Professional Social Science
Beverly Hills   London   New Delhi

Copyright © 1984 by Sage Publications, Inc.

For information address:

SAGE Publications, Inc.
275 South Beverly Drive
Beverly Hills, California 90212

SAGE Publications India Pvt. Ltd. SAGE Publications Ltd
M-32 Market   28 Banner Street
Greater Kailash I   London EC1Y 8QE
New Delhi 110 048 India   England

Printed in the United States of America

International Standard Book Number 0-8039-2545-X
International Standard Book Number 0-8039-2324-4 (pbk.)

Library of Congress Catalog Card No. L.C. 84-050798

SECOND PRINTING, 1985

CONTENTS

INTRODUCTION TO
THE SCIENTIFIC GEOGRAPHY SERIES

Scientific geography is one of the great traditions of contemporary geography. The scientific approach in geography, as elsewhere, involves the precise definition of variables and theoretical relationships that can be shown to be logically consistent. The theories are judged on the clarity of specification of their hypotheses and on their ability to be verified through statistical empirical analysis.

The study of scientific geography provides as much enjoyment and intellectual stimulation as does any subject in the university curriculum. Furthermore, scientific geography is also concerned with the demonstrated usefulness of the topic toward explanation, prediction, and prescription.

Although the empirical tradition in geography is centuries old, scientific geography could not mature until society came to appreciate the potential of the discipline and until computational methodology became commonplace. Today, there is widespread acceptance of computers, and people have become interested in space exploration, satellite technology, and general technological approaches to problems on our planet. With these prerequisites fulfilled, the infrastructure needed for the development of scientific geography is in place.

Scientific geography has demonstrated its capabilities in providing tools for analyzing and understanding geographic processes in both human and physical realms. It has also proven to be of interest to our sister disciplines, and is becoming increasingly recognized for its value to professionals in business and government.

The Scientific Geography Series will present the contributions of scientific geography in a unique manner. Each topic will be explained in a small book, or module. The introductory books are designed to reduce the barriers of learning; successive books at a more advanced level will follow the introductory modules to prepare the reader for contemporary developments in the field. The Scientific Geography Series begins with several important topics in human geography, followed by studies in other branches of scientific geography. The modules are intended to be used

5

as classroom texts and as reference books for researchers and professionals. Wherever possible, the series will emphasize practical utility and include real-world examples.

We are proud of the contributions of geography, and are proud in particular of the heritage of scientific geography. All branches of geography should have the opportunity to learn from one another; in the past, however, access to the contributions and the literature of scientific geography has been very limited. I beleive that those who have contributed significant research to topics in the field are best able to bring its contributions into focus. Thus, I would like to express my appreciation to the authors for their dedication in lending both their time and expertise, knowing that the benefits will by and large accrue not to themselves but to the discipline as a whole.

—Grant Ian Thrall
Series Editor

SERIES EDITOR'S INTRODUCTION

Central Place Theory seeks to provide an explanation of the numbers, sizes, and locations of urban settlements in essentially rural, farming regions. Why is it, for example, that there are few large cities, many more towns, and an even larger number of small villages or hamlets in such regions? Why is it that the smaller places are located closer together and the larger ones further apart? What are the relations between the roles of the different-sized urban settlements? How do these patterns and arrangements change over time and from one region to another? These are the sorts of questions addressed by central place theory.

The intellectual roots of central place theory can be found in the works of rural sociologists and geographers in the early 1900s, but the main contributions to the development of the theory were made in the 1930s and 1940s by two German scholars, Walter Christaller and August Lösch. In their studies, the economic interdependencies between town and country were spelled out, and the notions of a hierarchy of economic functions and a corresponding hierarchy of different-sized urban settlements were developed. The framework that they proposed has been found useful in interpreting settlement patterns, in explaining the decline of many small villages, in planning the location of new settlements and in analyzing the social structures of rural communities. The theory has attracted the interest of scholars not only in geography, but also in anthropology, economics, planning, and sociology.

Leslie King provides not only an overview of central place theory and its antecedents but also a description of the different lines of work that have flowed from the theory. For example, King presents in this book a commentary on the work of economists who have sought to use central place theory as the framework for an explanation of how centers of production are arranged. He also provides an important review of the large body of work done by archaeologists and anthropologists on the social structures of market systems in countries as dissimilar as China and Guatemala. He discusses the applications of central place theory in planning settlements in the vastly different contexts of Canada, Ghana, and Israel. Finally, King outlines some more recent attempts to rewrite the central place theory in more formal mathematical language.

Throughout this work, King keeps the discussion at a nonmathematical and nontechnical level while relying upon numerous diagrams and maps taken from various studies. This introduction to central place theory should be of particular interest to people whose interests lie in the fields of urban geography, economic geography, human ecology, rural sociology, regional archaeology and anthropology, regional planning, regional economics, or regional science.

—Grant Ian Thrall
Series Editor

CENTRAL PLACE THEORY

LESLIE J. KING
McMaster University

1. INTRODUCTION

Cities and towns differing in size and character are to be found in all parts of the settled world. In the regions that we know today as China, Egypt, Israel, Iraq, Pakistan, Greece, and Turkey, the histories of cities as important centers of human activity extend back thousands of years to the dawn of civilized life. Their origins in those ancient times may have been as religious and administrative centers or defensive sites. Other cities, in these same regions and in other parts of the world—in Europe, Asia, the Americas, and Africa—began as important locations for trade and commerce and in this role they have grown and flourished over the centuries.

The economic forces that have given rise to the development and continued growth of cities are numerous and complex. At the most basic level they have had to do with the development of agriculture and trade and the emancipation of a large proportion of the population from the daily tasks of providing food and subsistence. Freed from these subsistence activities but dependent on enjoying the products of other men's labor, the city dwellers have been able to develop trade and commerce and all of the financial and political arrangements needed to support those activities. With the subsequent accumulation of wealth within the cities there came also the development and patronage of the arts and cultural activities.

The economic rise of the cities was inextricably tied to the development of capitalism, as the French historian Fernand Braudel (1982) has portrayed so vividly. The profit motive and the quest for capital accumulation drove the economic machine that was the city, and as success followed upon success the city grew and prospered.

9

The particular economic advantages of locating businesses and industries together in the clustered form of the city are referred to by economists as *agglomeration economics*. The term is a general one and encompasses those benefits, for example, gained from locating near one's suppliers or markets; those realized from being in touch with other firms in the same industry; those accruing from the sharing with other industries of city services such as banking, insurance, utilities, and so on; and those associated simply with being in a city and enjoying access to its labor force, its markets, and its amenities. Under capitalism, the advantages of clustering economic activity within the cities were highlighted and promoted. The universality of these benefits, however, is attested to by the fact that even today in countries that denounce the capitalist way of life, the cities remain the principal centers of economic activity.

Today, in fact, most of the world's population lives in cities, towns, and villages of varying sizes. The term *urban* is used to distinguish this larger proportion of the population from the smaller *rural* population, and Table 1.1 shows how the urban proportions have changed over the past half century and how they are predicted to change in the future. In every major region of the world the urban population has been increasing and at an especially fast rate in the less developed regions. It is convenient to use the words *urban place* or *urban center* to refer to any city, town, or village, whatever its size. It is not, of course, the size of the center's population alone that distinguishes it from a rural community. There are villages, for example, in many parts of the less developed world that may have many hundreds or even thousands of inhabitants who are engaged principally in the subsistence activities of farming, hunting, and fishing. These villages would not be thought of as urban centers in the sense that *urban* implies a way of life consisting of types of economic activity and patterns of social and cultural organization that are radically different from those of such communities. Above all, the urban center is engaged in levels of commerce and industry that are far removed from the basic subsistence activities of the rural village.

The point at which a community is properly described as urban rather than rural is not easily determined, therefore, and countries around the world use different population sizes to describe what is urban and what is not. In the United States it has generally been a population of 2,500 persons or more that has qualified a center as urban; in Canada and Australia the threshold has been set at 1,000 persons; while in Denmark and Sweden the cutoff is even lower at 200 persons. In Austria and Belgium, by contrast, a center must have a 5,000 inhabitants to qualify as urban, and in Japan only those with populations over 20,000 are considered urban!

TABLE 1.1 Percentage of Total Population in Urban Localities in the World and Eight Major Areas, 1925-2025

Major Area	1925	1950	1975	2000	2025
World total	21	28	39	50	63
Northern America	54	64	77	86	93
Europe	48	55	67	79	88
USSR	18	39	61	76	87
East Asia	10	15	30	46	63
Latin America	25	41	60	74	85
Africa	8	13	24	37	54
South Asia	9	15	23	35	51
Oceania	54	65	71	77	87

SOURCE: United Nations (1974, p. 63).

The difficulty involved in distinguishing smaller urban centers from what are essentially rural communities is in many ways a reflection of the close relationships that exist between the two communities. In economic terms, the urban place provides the market center for the farmers; in turn, the merchants in the urban place are dependent, in part, upon trade with the farmers. Intertwined with these economic relationships are many other cultural, political, and social ones. Over the centuries these relationships between the urban center and the surrounding countryside often have been masked by the more complex and longer-distance relations that have developed between villages and towns, towns and cities, and cities themselves. But whether as the hub of an overseas commercial empire (as has been the case, for example, with Amsterdam, Athens, Lisbon, London, Madrid, Marseilles, New York, Paris, and Venice), or as the center of a regional trading network (such as Chicago, Leningrad, Munich, Peking, or Prague), or simply as the leading commercial center within a large region (for example, Birmingham in the United Kingdom, Lyons in France, or St. Louis in the United States), the city has always been dependent, and remains so, upon a set of very complex relationships with the surrounding rural countryside. Some have described this interdependency as a parasitic one in the sense that the urban center exists upon and collects the surplus (both agricultural and monetary) that originates within the rural community. Others have described the relationship as one of dominance in the sense that there is "no town that does not supply its hinterland with the amenities of its market, the use of its shops, its weights and measures, its money-lending, its lawyers, even its distractions" (Braudel 1982, p. 374). There are elements of truth in both views, and this volume seeks to tease apart these elements in order to discover the relationships that exist not only between village and countryside but between village and town, town and city.

The overwhelming dominance of the larger cities in the settlement patterns of the world is clear for all to see. The numbers in Table 1.2 simply serve to emphasize the point. They show that the proportion of the world's population living in cities of one million inhabitants or more has increased from 9 percent in 1960 to 13 percent in 1975; and although there are signs in the industrialized nations that this concentration is slowing down and that the smaller urban places are experiencing an economic revival of sorts, there is no reason to suppose that the trend will be reversed in any significant way at the international level.

Below these large metropolitan areas there are within any country innumerable smaller cities, towns, and villages that are urban in the sense that they are centers of industry and commerce, of art and culture, and of political, economic, and social power that distinguishes them from the rural countryside. It is these towns, villages, and, as the smallest centers often are called, *hamlets*, that together form the principal subject of this book. In these smaller urban centers the interrelationships between town and country, between urban and rural society, are clearly exhibited; at this level they can be studied, and generalizations can be made.

In an earlier paragraph, the concept of agglomeration economies was introduced in describing the strong tendency for economic activities to cluster together in the particular locations occupied by cities or other urban places. It was noted also that particular forms of agglomeration economies might apply in different situations and to different activities. The location, for example, of an automobile parts manufacturer in a city would, undoubtedly, be an expression of a different economic benefit than would be the case with the location of a farm machinery dealership in the same center. In the first case, the locational attraction might be the nearby presence of a steel mill for its supplies or an automobile manufacturer for its market, or both; but in the case of the second, the farm machinery dealership, the attraction might be the market represented by the surrounding farming community and the presence in the city of appropriate financial support.

The farm machinery dealership case involves an element of what is called *centrality*. For many economic activities located in a city, town, or village it is important that they be central to some larger market area that extends beyond the boundaries of the urban place itself. This is especially so for activities that serve the farming community surrounding the urban place.

This notion of centrality is one that must be pursued with care and not too much force. In a hypothetical sense, it is obvious that any businessper-

TABLE 1.2 Number and Population of Million-Cities, and Percentage of Total Population in Million-Cities, 1960 and 1975, in the World and Major Areas

Area	Number of Million-Cities		Population of Million-Cities (millions)		Percentage of Total Population Million-Cities	
	1960	1975	1960	1975	1960	1975
World total	109	191	272	516	9.1	12.8
More developed regions	64	90	173	251	17.7	21.9
Less developed regions	45	101	99	265	4.9	9.2
Europe	31	37	73	93	17.3	19.3
USSR	5	12	13	25	6.1	9.7
Northern America	18	30	52	80	26.2	32.9
Oceania	2	2	4	6	24.7	26.9
South Asia	16	34	32	88	3.7	6.8
East Asia	23	45	60	131	7.7	12.9
Africa	3	10	6	22	2.4	5.5
Latin America	11	21	31	71	14.5	21.9

SOURCE: United Nations (1974, p. 36)

son seeking to gain maximum advantage (and profit) would locate as close as possible to as many potential customers as possible, and in the realm of theory this would require that he or she locate at the center of the market area. But in the real world, the feasible locations for the business may be quite off center to any potential market area and, besides this area itself is not predetermined and will be shaped subsequently by the actions of the businessperson. These cautions aside, it has proven useful, as will be illustrated in this book, to think of an urban center located in a farming region as a central place offering a number of business activities to the rural population located in the areas around it. The term *central place* actually was coined back in 1931 when the American geographer, Mark Jefferson, wrote; "Cities do not grow up of themselves, countrysides set them up to do tasks that must be performed in central places" (1931, p. 454).

This book is about central places, about economic activities that they offer, about their comparative sizes and their relative locations, and their changing fortunes over time.

The discussion begins with some further observations on the character of urban places in different parts of the world, emphasizing their comparative sizes, their functions, and their trade areas. The interest at this point is on noting how these features have been examined and described in a number of studies of North American and European urban settlements.

The study of urban settlement patterns has produced its share of generalizations and even attempts to fashion these generalizations into broader conceptual schemes or theories. The "central place theory" of Walter Christaller and August Lösch, two eminent German economic geographers, is one well-known schema that has been the subject of much additional discussion. It is outlined in Chapter 3 of this book.

There have been many criticisms levelled at central place theory—many of them have been philosophical in nature, while others have been more concerned with the lack of historical and empirical support for the theory. Central place theory is dismissed by many as being reflective of certain existing social patterns at particular times in history and therefore incapable of yielding any powerful predictions that might be tested in different contexts and at different times. Nor can central place theory suggest, continue the critics, ways of refashioning society that might improve the human condition for all concerned. It must be conceded that these criticisms have force; but nevertheless, there have been ways in which central place theory has stimulated further creative thought about urban processes and patterns and has provided useful guidelines for the planning of urban settlements. A number of these contributions are reviewed in Chapter 4 of this volume.

Creativity in human thought is often stimulated by dissatisfaction with existing explanations or conceptualizations. The contentions that central place theory is formulated imprecisely in regard to its treatment of economic production, of shopping behavior, of the locational arrangement of urban places, and of the process of urban growth have been the starting points for many new attempts at theory writing, some of which are the subject of the final chapter of this book.

2. CITIES, TOWNS, AND VILLAGES

It was noted in the introduction that there is a wide range in the population sizes of the urban centers in any country and the example of the United States may be used to illustrate this point. In Table 2.1 the changing numbers of urban places in the United States are shown for the period 1900 to 1980. Although the population size intervals used in the table are not constant, nevertheless the data show that the numbers of places in all size categories have increased, except for the two largest sizes over the past decade.

Urban centers differ not only in their size but in their functions. What is meant by function? There are various types of economic activities carried on in any urban center, such as manufacturing, wholesaling, retailing, and many forms of personal and business services. In any one urban place, for any number of reasons, some of these forms of economic activity may be more important than others. A village is located near a coal mine, a town built on the site of mineral springs, or a city specializing in the production of iron and steel are obvious examples of urban centers that have specialized economic activities located within them. Other centers may emphasize the offering of retail activities (food, beverages, clothing, and so on) and services (banking, insurance, barbers, churches, and so on) to people living within the centers and, more importantly, to the farming population living around the center. These dominant activities of an urban center serve to define its function. In this discussion, the emphasis will be placed on urban centers that exist primarily to serve the inhabitants of the surrounding rural areas. This is their principal function and, as was noted in the introduction, they are referred to as central places. In the second and third subsections, attention is given to the nature of central place functions and to the character and extent of the trading areas of urban places.

Population Size Considerations

Many writers have discussed the idea of a relationship between the size of an urban place and its rank among all of the urban places in terms of size. For the United States when the urban centers are plotted on a graph according to their size and rank (as in Figure 2.1) there is a fairly regular progression that could be approximated or described by a curve. By contrast, in the case of Mexico, for example, the progression is not as regular and the largest city is some considerable distance apart, at least on the graph, from the majority of the other centers (Figure 2.1).

What do these observations suggest? For one thing, the question of the relationship, if any, between size and rank has prompted a number of attempts to provide more precise mathematical statements or models of the relationship and to explain why the fit is not as close in some countries as in others. The so-called "rank-size rule" is one product of this interest. The rule is written usually as, $p_i = k/r_i$, where p_i is the population of a city, designated as i, r is its rank in size, and k is the population of the largest city.

In the early twentieth century, the rank-size rule was seen as a useful method of describing the urban size distribution of the United States, but it is now no longer considered applicable or useful. The contention that

16

TABLE 2.1 Number of Urban Places in the United States, 1900-1980

Class and Size of Population	1900	1940	1970	1980
Under 2,000	*	*	627	1,016
2,500–5,000	832	1,422	2,295	2,665
5,000–10,000	465	965	1,839	2,181
10,000–25,000	280	665	1,385	1,765
25,000–50,000	82	213	520	675
50,000–100,000	40	107	240	290
100,000–250,000	23	55	100	117
250,000–500,000	9	23	30	34
500,000–1,000,000	3	9	20	16
Over 1,000,000	3	5	6	6
Total number	1,737	3,464	7,062	8,765

SOURCE: United States Bureau of the Census (1960, 1970, 1980)
*Data unavailable.

there may be a regular and uniform progression in population size of urban centers from the smallest up to the largest in any country is not only difficult to reconcile with the statistical facts in many parts of the world but it also runs counter to those established traditions in literature, social science, and everyday life that emphasize the distinctions between cities as large places, towns as intermediate-sized centers, and villages and hamlets as small places. In other words, people are accustomed to using the words "cities," "towns," and "villages" to refer to categories of urban centers without worrying about precise definitions being attached to these categories in terms of their population sizes. In everyday conversations there are references made to "city hall," to the "bright lights of the city," to "town and gown" relations, to "village communities," and so on. The speaker usually has a good sense of the type of urban community to which he or she is referring when such terms are used, but there may be little uniformity from one view to another in regard to the population size of the communities involved. The resident of San Francisco obviously will have a different notion of a city and of city life from that of the resident of Columbus, Ohio!

In the same manner, many reports in social science use similar categories. Consider, for example, the following passage that appeared in a 1974 report on metropolitan America: "Although small towns and densely packed cities still persist and even though an integrated national system of settlements

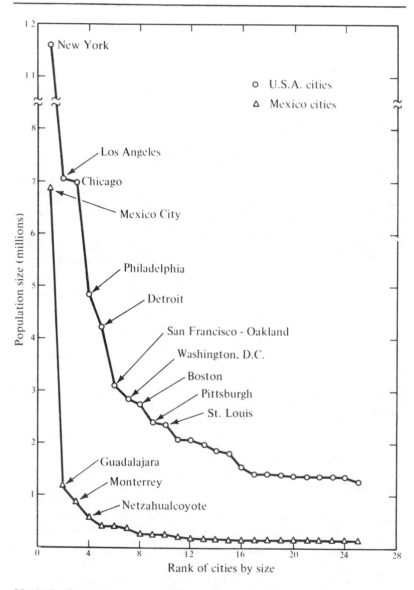

SOURCE: King and Golledge (1978)

Figure 2.1 Plots of Rank-Size Urban Distributions for the United States and Mexico, 1970

may emerge in the future, urbanization in the twentieth century has been dominated by the rise of metropolitan communities" (Hawley et al. 1974, p. 7).

The process of taking and repeating national censuses does require the establishment of certain definitions of what constitutes a particular type of urban place. Some such definitions were noted in the introductory chapter. In most countries, the United States and Canada included, an "incorporated" urban place or "municipality" is one that has been legally recognized as a political subdivision with a defined area and with certain powers of government. These powers generally include those of levying some form of taxation, typically on property, and of administering the provision of services (water, sewerage, fire, police, and so on). In the United States an urban center may be incorporated as a city, borough, town, or village depending on its size and the provisions of the law under which it is incorporated.

Most national censuses define the largest urban places in very precise terms. In the United States, a Standard Metropolitan Statistical Area (SMSA) includes at least one central city of 50,000 persons or more, the county in which this central city is located, and those adjoining counties that are metropolitan in character. The determination of what is metropolitan relies on (1) employment characteristics—at least 75 percent of the county labor force must be employed in non-agricultural activities; (2) population density; and (3) the extent of economic and social interaction of the county with the central city county (as measured, for example, by the percentage of the county's workers working in the central county). The 1970 census recognized as many as 243 SMSAs, which are shown in the map in Figure 2.2; by 1980 the number had increased to 318.

These SMSAs in 1980 covered an area of more than 566,000 square miles, which was approximately 16 percent of the land area of the United States. The population living within them in 1980 was over 169 million people (almost 75 percent of the total United States population), which is a reminder of the importance of the very large centers in the urban pattern of most developed countries, as stated earlier. This population, however, was not evenly spread over the SMSA areas but was was heavily concentrated within the more restricted urbanized areas that accounted for only slightly less than 10 percent of the total SMSA land area but more than 80 percent of the SMSA population. The impression created by the map in Figure 2.2 that all of the SMSAs are given over to urban uses tends, therefore, to be an exaggeration as a great deal of this land is still rural.

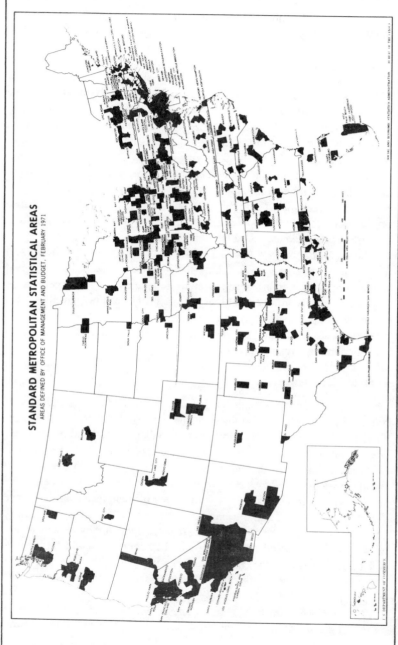

Figure 2.2 Standard Metropolitan Statistical Areas: United States, 1970

The fact that the metropolitan areas are defined so precisely in the census does not imply that the other classes of urban places are similarly well defined. The convention—and this is true of national censuses of other countries as well as of the United States—is to group and describe the urban places by way of arbitrarily defined population-size classes. Hence, in Table 2.2 it is seen that there were 117 places of 100,000 to 250,000 size in the United States in 1980, 290 places of 50,000 to 100,000, and so on. These groupings by way of population size are convenient for demographic analyses but convey no information about the economic or social character of the places themselves. The following sections shall refer to different-sized classes of urban places but shall attach added significance to these classes by referring to the characteristic types of economic activities or urban functions associated with the different levels. There will be a reliance on functions as well as population size in differentiating the hamlet from the village, the village from the town, the town from the city, and so on.

The premise that underlies the following discussion is that the idea of a continuous rank-size relationship—the idea that there is a smooth and regular progression in urban size from the smallest up to the largest center —is not a very useful one, and that greater insight can be gained into the nature of an urban pattern by referring to discrete population-size classes. The emphasis will be on studying a hierarchy of urban places rather than a continuous-size distribution of them.

Central Place Functions

The role of urban places as service centers for the farming regions surrounding them largely determines the character of the major economic activities carried on within the city, town, or village. They will be activities whose continued economic well-being in the urban center is largely dependent on the location of the place somewhere central to the region of farm population demand. These *central place functions* will generally be service activities and will not include those brances of manufacturing that serve more distant markets and are unrelated to the needs of the rural region. This is not to say that central locations are not sought after by such manufacturing activities, for indeed they are; but their location in an urban center is not an expression of the *complementarity* in function that exists between that place as a central place and the surrounding rural region and its inhabitants.

A central place function, then, is any activity carried on in the urban place that derives at least part of its support from people living in the rural areas around the place. Down through history, their character has been essentially the same. Consider, for example, the following description:

> At the bottom of the hierarchy in early modern western Europe even small villages had butchers, carpenters, shoemakers and perhaps a miller and a minister. Larger villages and towns boasted in addition bakers, innkeepers, a barber, a schoolmaster, perhaps a lawyer and a doctor, weavers, a clothier. The county town or provincial city had a far wider range of those engaged in dealing and administering, representatives of the upper echelons of the church and the professions and finer subdivisions of crafts and trades which appear undifferentiated in smaller centers. A major regional center, great port, or *a fortiori* a capital city took the process of specialization a stage further, once more paralleling the degree of differentiation to be found lower down the hierarchy but adding new titles to the list of distinctive forms of livelihood which could find a demand for their services. Characteristically this demand, though most intense within the great city, did not arise exclusively within it. An enclosure negotiation in Lincolnshire might find its way eventually to the Court of Chancery. Spitalfield silk goods commanded a market in Westmoreland as well as Westminster. Goods and service flowed in both directions through the channels of trade and communications.'' (Wrigley 1978, p. 300)

In many studies of central places, as will be noted later, the list of such functions has approached one hundred or so. The list typically includes such activities as ''grocery and food provision,'' ''laundering,'' ''sale of hardware,'' ''sale of clothes,'' ''doctors,'' ''dentists,'' and so on. Often the list is expanded to include social and cultural activities such as schools, churches, and clubs.

In most urban places any one central function will generally be offered by more than one business or unit. There may be, for example, many foodstores, several doctors and dentists, and two or three churches. Each business unit offering a central function is counted as one *functional unit*. Obviously, the number of these must at least equal the number of central functions and it will usually be greater.

Furthermore, any one building or establishment in the urban center may involve more than one function and functional unit. For example, a general store that sells mainly food might also have a hardware section and serve as a post office. In this case, there is one establishment, three functions (sale of food, sale of hardware, post office), and three functional units. This distinction between establishment and functional unit is important in certain cases, but on the whole it does not warrant much emphasis.

22

TABLE 2.2 Relationship Between Urban Population Size and Number of Urban Functions in Different Areas

Study	Correlation Coefficient, r
Iowa[a]	0.890
Canterbury, New Zealand[b]	0.823
South Illinois[c]	0.892
Barrie area, Ontario[d]	0.870
Wales[e]	0.87

a. Berry, Barnum, and Tennant (1962)
b. King (1962)
c. Stafford (1963)
d. Marshall (1969)
e. Carter, Stafford, and Gilbert (1970)

What relationships should exist between these particular features of an urban place and also between them and its population size? First, for a group of urban places the number of functions and population size should be positively related: that is to say, the larger centers should offer more functions, the smaller ones fewer. The graph of this relationship should look something like Figure 2.3. The reasoning is that more people will visit the larger centers and this greater "pulling power" combined with the larger population in the town itself will enable certain more specialized functions to be offered there that could not be offered economically in the smaller places. Specialized medical clinics, for example, can be found in larger towns, but are absent in the very small urban places.

This positive relationship between size and number of functions has been confirmed in a number of studies for different areas all around the world. Table 2.2 summarizes some of these studies. The correlation coefficient obtained in each case for the number of functions/population-size relationship confirmed a strong positive one.

The graph in Figure 2.3 can be used to illustrate some possible anomalies. Points A, B, C, and D can be thought of as hypothetical urban places with particular population sizes and numbers of functions. The places represented by points A and B obviously have more functions than is suggested by the curve. A might be a small center with a very advantageous location—for example, at the junction of two major highways—that, on the basis of a high volume of through-traffic, could support more functions than might otherwise have been the case. City B might also be similarly located, or it might be an important regional center comparatively isolated from other towns of a similar size and therefore be able to offer a wider range of func-

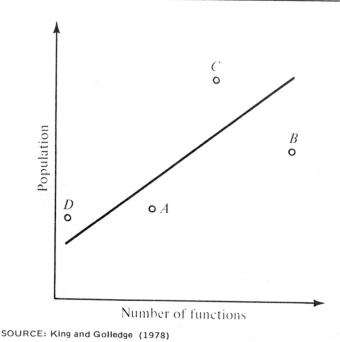

SOURCE: King and Golledge (1978)

Figure 2.3 Hypothetical Relationship Between Urban Population Size and
the Number of Functions in Central Places

tions. By contrast, places C and D have fewer functions than are suggested
by the graph. Place C might be a suburban community located fairly close
to a larger city and cast, therefore, in the role of a residential community.
Considering that persons living there presumably rely for the provision of
goods and services upon the larger city in which many of them work, place
C does not need to offer them. Place D, however, could be a small hamlet
that has suffered from improved transportation systems that now enable people to bypass it on their way to larger centers. People—most probably older
people—might still live there, but many of the center's economic functions
may have disappeared.

What of the relationship between the number of functional units and
population size? In this case, a nonlinear relationship such as is shown in
Figure 2.4 might be expected. This suggests that as population size increases,

24

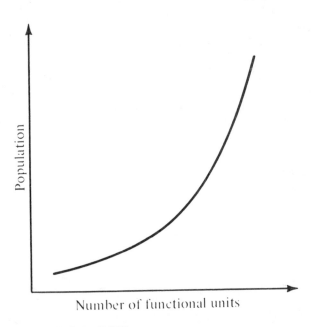

SOURCE: King and Golledge (1978)

Figure 2.4   Hypothetical Relationship Between Urban Population Size and the Number of Functional Units in Central Places

the number of functional units increases fairly rapidly up to a point but at a slower rate thereafter. This would be consistent with the notion that in the larger centers economies of size become feasible, thereby allowing larger operating units to be substituted for several smaller ones. The point is illustrated in the case of foodstores: In large centers big supermarkets replace the separate foodstores that are more typical in smaller urban places.

Again, deviations from this generalized relationship could be explained in terms of the town's particular location and/or the presence of functions other than central ones. Studies have confirmed that this relationship between size and the number of functional units is also a strong positive one.

This last relationship has provided a basis for estimating the *threshold* level for a function. The threshold level is defined as the minimum level of support, as measured by numbers of population, required to support a function in a particular community. For example, if there were five hairdressers serving a total population of 8,000 in both the town itself and its surrounding region, then a crude estimate of the threshold level for this function might simply be 8,000/5 = 1,600 persons. However, because this ratio is certain to vary from town to town, some form of statistical averaging is preferable. We shall refer later to how this is done.

Trade Areas

The description and delimitation of the trading areas of urban centers have long been subjects of interest to historians, geographers, and sociologists.

Braudel, in his book *The Wheels of Commerce* (1982), provided an extremely colorful account of how the trading realms of the towns and cities developed in Europe and elsewhere in the period from the fifteenth to the eighteenth centuries. For Braudel, the "instruments of exchange" whereby trade and commerce developed in the urban center were the *market,* "with all the bustle and mess, the cries, strong smells and fresh produce" (p. 28), the *shop* "as a fixed point of sale" (p. 71), traveling *peddlers* "who carried on their backs their very meagre stock" (p. 75), the *fair* which served "to interrupt the tight circle of everyday exchanges" (p. 82), and the *exchanges* and *stock markets* where the higher forms of financing and trading were localized. These were the building blocks of the trading networks and empires that emerged around particular towns and cities. Braudel observed that

> A town or city lies at the centre of a number of interlocking catchment areas: there is the circle from which it obtains its supplies; the circle in which its currency, weights and measures are used; the circle from which its craftsmen and new bourgeois come; the circle of credit (the widest one); the circle of its sales and the circle of its purchases; and successive circles through which news reaching or leaving the town travels. Like the merchants shop or warehouse, the town occupies an economic area assigned it by its situation, its wealth and its long-term context. (p. 188)

As an example of this pattern, Braudel cited sixteenth-century Nuremberg, Germany. Nuremberg, located at "virtually the geometric centre of the economic life of Europe during the early sixteenth century" (p. 188), had trading connections to the Middle East, India, Africa, and the New World; the global-scale catchment areas overlaid the city's local catchment areas in Germany.

The trading connections of not just one city but rather of groups or sets of urban places have been examined in even greater detail for many regions

26

throughout the world. Two examples are mentioned here—one from England, and the other from the United States.

Robert Dickinson (1947, p. 87), a geographer, in describing the urban settlements in East Anglia of England around the middle of this century, suggested that there were four classes of settlement according to their functions. At the top were those centers with populations of 5,000 or more that had "general livestock markets with aggregate sales exceeding 10,000 head per annum, three or more banks, a cinema, newspaper, secondary school and usually some local industries." Below them were the centers of 1,500 to 5,000 persons that had similar functions but were experiencing declines in population and market size, mainly because of the development of rail services. A third group of settlements, all of 1,000 to 2,000 persons, had thriving livestock markets but "no cinema, newspaper, etc., and only one or two banks." Finally, at the lowest level, there were the centers with populations of 1,000 to 2,000 that had no livestock markets and only one or two banks and specialized retail services.

Dickinson's classification highlighted the role of the urban settlements as the market centers for the farm output, specifically lifestock. In studying the functions of urban settlements today, it is easy enough to overlook this role as a market center given the wide range of other functions present that provide many different goods and services. But it is in the presence of a livestock or agricultural produce market that the fundamental interdependency between town and countryside is seen most clearly. Both Braudel and Dickinson have reminded us of this fact.

In the United States, the relationship of towns and trade areas received considerable attention during the first two decades of the present century from sociologists at the University of Wisconsin. The greater part of this work was begun by C. J. Galpin (1915) who, in his studies of Wisconsin rural communities, introduced several of the concepts that were to form the basis of much of the later work in the field. Galpin recognized that urban centers offering the same level of services should be equally spaced and that, theoretically, the trade area of any town should be circular (although in reality the trade areas of adjacent towns may overlap). A hexagonal pattern of trade areas was implicit in Galpin's formulation: He realized that if all of the farm territory were to be served, then it would be essential for the towns to be arranged in such a way that only six centers were at equal distance from any one center. However, no attempt was made to elaborate on the details of such a locational pattern.

Galpin's work was continued and enlarged upon by J. H. Kolb (1923, 1940), who developed one of the earliest generalizations of a set of trade

27

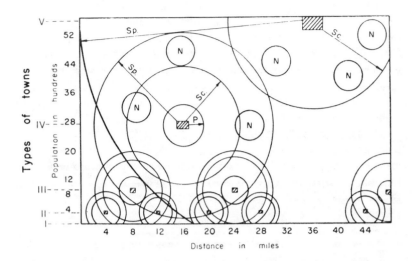

SOURCE: Adapted from Kolb and de Brunner (1940).
NOTE: Town types: I—hamlet; II—small village; III—village or small town; IV—town or small city; V—city. The primary service area has radius P; the secondary service area Sc; the specialized service area Sp; and the country neighborhood area is N.

Figure 2.5 Hypothetical System of Central Places

areas. Again, as in Galpin's studies, the functional relationship between town and country was the focal point of Kolb's research. In his words, "every farmer's gate opens onto a road and that road leads to a village or town. Because of this web of communication and transportation, there is an intermesh of town and country relationships" (Kolb, 1940, p. 109). Kolb defined three main tributary areas, at least two of which are recognizable for any urban settlement. There was, first of all, the *primary* or personal service zone. This was a very small zone and encompassed those goods and services that were most frequently demanded by the farm population. Kolb insisted that such a zone could not be distinguished around the larger cities. By contrast the *secondary* service zone existed for all urban centers and it represented the generalized trade area for the greater number of the goods and services offered by any town. The third tributary zone was the *specialized* service area, which the smaller urban settlements did not possess because they did not offer the more specialized type of services such as technical and university education, medical specialists and hospital services, and high quality retail establishments.

The urban centers associated with Kolb's settlement model were divided into five main types ranging in size of population from the 200 or so inhabitants of the small hamlet up to 5,000 or more persons for the city. The manner in which these towns were spaced relative to one another in an idealized pattern is shown in Figure 2.5. This "model" settlement pattern was a noteworthy forerunner of the more theoretical frameworks developed later by scholars in Europe.

The settlement patterns in the United States, especially in the Midwest, in the early decades of this century were comparatively recent ones. Nevertheless, a well-defined hierarchy of urban centers was seen to be emerging; and in the studies by Galpin, Kolb, and others there were important generalizations drawn about these trends.

3. A THEORY OF CENTRAL PLACES

The studies of settlement patterns and trading areas made by geographers, and sociologists in particular, suggested that there were generalizations that could be made about these patterns. The contribution by Kolb of an idealized location pattern of market towns was one attempt. The extension of this approach to acknowledging the possibility that laws and theories of settlement might be discovered—similar in form and function to the theories being developed in the physical sciences—was an attractive proposition to many scholars in the early to middle decades of this century. Even today, the prospect remains an appealing one for some although the enthusiasm for emulating the approach of the physical sciences has waned considerably.

In this section, the work of two eminent German scholars is reviewed. Both of them sought to formulate theories of urban settlement in which central places played the preeminent role. Their contributions were seminal in the sense that from them grew an extensive literature on central place theory and central place systems. The bibliography published by Berry and Pred (1961) provides signposts to this literature. The books by Berry (1967) and Beavon (1977) provide excellent accounts of the theories.

Walter Christaller's Theory

Christaller published his work in German in 1933. He introduced his discussion with the question, "Are there laws which determine the size, number and distribution of central places?'' He believed that there were such laws to be discovered and that logic could be used to weave them together into a theory. This theory, in turn, could be tested and verified with observations on the urban settlement pattern of the Southern Germany of his time.

The cornerstone of Christaller's theory was the idea of a functional interdependence between a town and the surrounding rural area. This was by no means an innovation in the fields of settlement studies and rural sociology, as has been noted already, but Christaller formalized the notion in a decidedly new way. Upon the basic premise that "the chief profession, or chief characteristic, of a town is to be the center of a region" (Baskin 1966, p. 116), he constructed a completely new framework for the study of settlement geography. Christaller did not ignore the fact that in contrast to central places per se, there exist various other types of settlements—for example, the "pointly bounded places" such as agricultural villages, or the "areally bounded places" which include mining towns, bridge and fortress towns, harbors and ports, border and custom towns—but these other places were disregarded in his discussion. The focal point of Christaller's attention was the central place with its central goods and services.

In the development of any theory of the real world it is always necessary to make certain limiting assumptions in order to be able to focus on those features that are of interest. Thus, for example, in attempting to develop a theory of central places that emphasizes the economic interrelationships between them and the rural areas, it would be inappropriate to include such features as the importance of a defensible site or the existence of mineral springs thought to have healing powers as the reasons for the locations of particular towns. These factors might be important in other discussions, but they were irrelevant to the development of Christaller's theory and were ruled out by a number of his assumptions.

ASSUMPTIONS

Christaller assumed first of all that there was a boundless and homogeneous plain with soil fertility and other natural resources being the same in all parts of it. This plain was settled uniformly, and the farmers everywhere had the same levels of income and the same demand for goods and services. Travel across the plain was equally possible in all directions, and the costs of travel and of transporting goods were a function only of the distance travelled.

Christaller assumed further that both the farmers as consumers and the businesspersons in the urban places as the producers of goods and services were rational individuals who would seek to minimize their costs (whether they were transport or production costs) and to maximize their profits. From the point of view of the consumers, this would imply that they would travel only to the nearest central place that provided the goods and services that they demanded.

On the part of the businesspersons it meant that a good or service would not be produced and sold if a profit could not be realized. If there was insufficient demand, for example, for them to at least break even, then it was assumed that they would not offer the service or produce the good. One further assumption that was made by Christaller that is related in part to the assumption of rational behavior and also to the assumption that new businesses could start up wherever and whenever they pleased) was that all of the settled plain would be equally well served by central places.

With these simplifications, the task for Christaller was then to describe a central place pattern that contained the minimum number of urban places and satisfied the different economic and behavioral conditions.

RANGE OF A CENTRAL PLACE FUNCTION

It was necessary, before proceeding to outline his theory, for Christaller to introduce and define one or two further concepts. The most important of these was that of the *range* of a central place function. This range had both an upper and lower limit. The upper limit, as Saey (1973) has stressed, was the key concept in Christaller's formulation of the hexagonal pattern of market areas and the hierarchy of central places; the upper limit was defined simply as ''the farthest distance the dispersed population is willing to go in order to buy a good offered at a place—a central place'' (Baskin 1966, p. 22). The more expensive the good, the greater the willingness to travel longer distances and hence the upper-limit range would be larger. For more frequently demanded goods, which would be the cheaper goods, the upper-limit range would be smaller. Although he emphasized the demand side in defining the notion of the upper limit to the range of a good, Christaller did acknowledge that the economics of the supply side also would affect the range. He noted, for example, that if a business required a given number of customer visits to break even, then it would have to extend the range of its business activity far enough out into the countryside to capture this level of demand. This lower limit of the range could be thought of as "the minimum amount of consumption of this central good needed to pay for the production or offering of the central good" (Baskin 1966, p. 54). Later, when he came to consider real-world deviations from the theory, Christaller would acknowledge that the density and distribution of population in an actual region, as opposed to the hypothetical world of the theory, would affect the range. In the same way, the fact that large central places offer greater opportunities for multiple-purpose shopping trips than do the smaller places may result in prices for a particular good being lower in the larger places than in the smaller. Hence, the range of a good in a large place may be greater than it is in the smaller place.

The threshold value for a central place function (which we touched on briefly in the previous chapter) is, of course, directly related to the lower limit of the range. It is a measure of the minimum level of demand needed to ensure that the offering of a good or service will be profitable. Again, Saey (1973) has pointed out that several writers on the subject have misinterpreted this concept of threshold as the principal one in Christaller's schema, which it is not. The key concept—the upper limit to the range—is ideal in that it is the maximum distance over which a good will be demanded; but in the case where there is another central place nearby that offers the same good, then there is a point at which it becomes cheaper for the purchaser to go to this other center. That point defines the *real* range of a good. The distinction between the two concepts is illustrated in Figure 3.1 where the ideal range is shown as s_2 and the real range as s_3. Finally, it is convenient to refer to the *order* of a place and/or central place function with reference to the size of its range. Lower-order places and functions have smaller ranges, both ideal and real, than do higher-order places and functions.

HIERARCHY OF CENTRAL PLACES

The range of a good or service establishes the size of the market area necessary to provide the economic support for the business offering the good or service. The larger the range, the larger the tributary area needed; hence, the business in question will be located in one of the larger urban centers.

In developing his argument, Christaller assumed that a well-developed urban system with one large city, a smaller number of towns, and a large number of villages and hamlets already existed in his hypothetical region. In other words, he did not seek to explain or describe how this hierarchy of centers came into existence or why some centers grew larger than did others. He simply took for granted that there was a hierarchical set of ur-

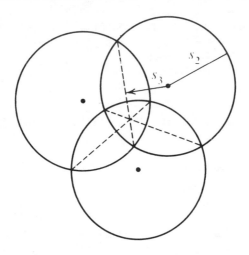

Figure 3.1 Ideal and Real Ranges of a Central Place Function

ban places and then used this as the basis for describing the arrangement of market areas and the number and distribution of urban centers.

The larger urban places would have the larger tributary or trading areas and they would be able, therefore, to offer those goods and services that could not be supported by the smaller urban centers. The larger places would be able, of course, to also offer everything that could be offered in a smaller place.

There would then be a hierarchy of urban places differentiated not only by their size but by the number and order of the functions offered by them. In his own study of Southern Germany, Christaller recognized seven main levels in the hierarchy (Table 3.1). As will be noted later, there was one other level—that of the smallest H places—but this need not concern us at this point. The other information given in this Table 3.1 will become clearer as we proceed with the discussion.

LOCATION PATTERNS

Essential to the ideal or theoretical central place pattern that Christaller described are certain features of the locational arrangements of the farm population and urban places.

TABLE 3.1   The Central Place System of Christaller's Southern Germany

Type of Place	Number	Number of Tributary Areas	Distance Between Places (kms)
M	486	729	4.0
A	162	243	6.9
K	54	81	12.0
B	18	27	20.7
G	6	9	36.0
P	2	3	62.1
L	1	1	108.0
Total	729		

Recall that one of the assumptions made earlier was that the settlement pattern over the plain was a uniform one. This implied that the farmsteads were equidistant from one another. A square pattern does not meet this requirement, as the diagonal distance is greater than that of the side. It is only a hexagonal distribution of farmsteads over the whole plain that will satisfy the requirement.

In the same vein, Christaller assumed that the urban places of each level of his central place hierarchy would be uniformly distributed throughout the region, in other words the hamlets would be equidistant from one another, the villages equidistant from one another, and so on up the hierarchy. That is what is shown in Figure 3.2 for five of Christaller's levels. In the diagram the small solid dots may be thought of as the farmsteads.

The diagram serves also to illustrate the following points. The largest or highest-order central place has a large tributary area, the extent of which is determined by the average real range of the highest-order functions offered there. But that same center also offers all of the lower-order functions that have smaller ranges and, therefore, smaller tributary areas. There is then for the highest-order center and, indeed, for each center at any level of the hierarchy, a set of tributary areas of differing sizes nested within one another. Further, it is clear that within the larger tributary area there are located many lower-order urban places.

These properties, and others, of the hierarchical central place system shown in Figure 3.2 were described in detail by Christaller. His main results are summarized in the following paragraphs.

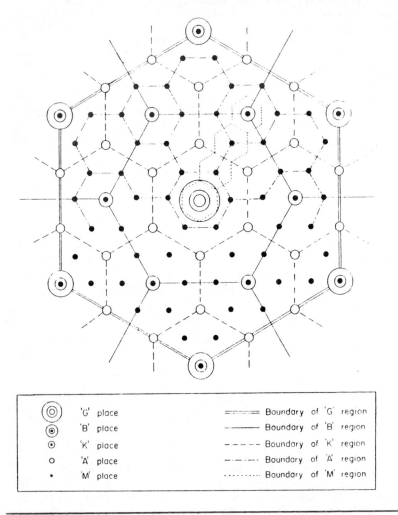

⊚	'G' place	══════ Boundary of 'G' region
⊙	'B' place	────── Boundary of 'B' region
⊙	'K' place	────── Boundary of 'K' region
o	'A' place	──·──· Boundary of 'A' region
•	'M' place	·········· Boundary of 'M' region

Figure 3.2 Market System of Central Places According to Christaller

NESTING OF TRIBUTARY AREAS

Christaller contended that under a pure marketing system—that is to say one in which all of the earlier assumptions held true—the pattern of central places shown in Figure 3.1 would prevail. In this pattern, it can be seen, there is always one center of a higher order surrounded by six centers of the next lower order and these six surrounding centers are located on the vertices of the higher-order center's largest tributary area. For example, in Figure 3.1 the one G place has six B places surrounding it and they are located at the vertices of the G region.

It is worth noting at this point that it is not essential that the relative location of centers be prescribed in this way and the assumption of a uniform distribution would still be satisfied if the centers were arranged as is shown in part b or c Figure 3.3. These options are not discussed further at this time but will be referred to again at a later point.

What does the arrangement of centers shown in Figures 3.2 and part a of Figure 3.3 imply about the degree of nesting? In Figure 3.2 there is obviously only *one* G region. But what of the B regions? There is one complete B region and another six B regions that are shared with other G places not shown on the diagram. Each G place serves one-third of each of these six B regions; this is the equivalent of two full B regions. Hence, there are the equivalent of three full B regions shown in Figure 3.2. Similarly, for the K places there are seven full K regions contained in the region plus one-third of six partial K regions, which adds up to a total of nine K regions. For the A places, the corresponding total is twenty-seven and for the M places is eighty-one.

This progression by threes in the number of tributary areas contained within the central place pattern is shown in column three of Table 3.1. The second column of Table 3.1 describes a similar but different progression in the number of places within the system. These numbers are explained as follows.

Each central place is presumed to perform all the functions of lower-order places. The highest-order place, therefore, also acts as a second-order place, a third-order place, and so on. Consider now what is meant by saying that "the equivalent of three second-order tributary areas exist within the highest-order tributary area." It means that in addition to the first-order place (now also considered as a second-order place) there are two other centers. Similarly, nine third-order tributary areas involve the first-order place, both second-order places all acting now as third-order places, and six other places. Hence, the progression describing the number of different-sized places in a system of degree three is: 1, 2, 6, 18, 54, 162, ... and so on.

THE SPACING OF CENTRAL PLACES

For the pattern shown in Figure 3.2 the distances between the different places can be calculated using simple trigonometry. The actual mathematical

36

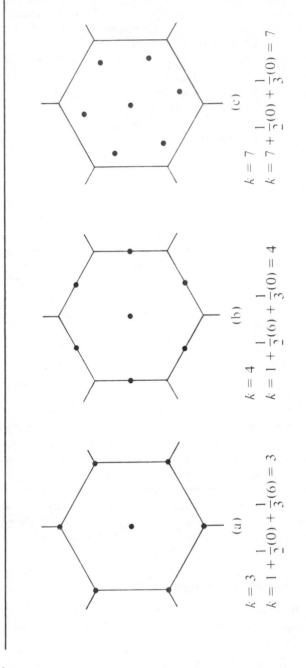

(a)

$k = 3$
$k = 1 + \frac{1}{2}(0) + \frac{1}{3}(6) = 3$

(b)

$k = 4$
$k = 1 + \frac{1}{2}(6) + \frac{1}{3}(0) = 4$

(c)

$k = 7$
$k = 7 + \frac{1}{2}(0) + \frac{1}{3}(0) = 7$

SOURCE: King and Golledge (1978)

Figure 3.3 Three Different Arrangements of Central Places

calculations are not presented here. The results show, however, that the distances between central places of progressively higher orders increase by a factor of $\sqrt{3}$. The actual distances are determined by the range of the lowest order place in the pattern. In Table 3.1 it can be seen that this basic distance, the range of the M region, was assumed to be 4 kms. Then, with this established, the distance between A places would be $4 \sqrt{3}$, which is 6.9 kms, that between K places would be $6.9 \sqrt{3}$, which is 12 kms, that between B places $12 \sqrt{3}$, which is 20.7 kms, and so on. These are the values given in the right-hand column of Table 3.1.

OTHER ARRANGEMENTS OF CENTRAL PLACES

Christaller insisted that the so-called marketing principle that underlies the pattern shown in Figure 3.2 would be the normal situation. He recognized, however, that other forces might distort the pattern and produce different central place arrangements. In particular, two competing principles—the one of traffic routing and the other of administrative partitioning—were recognized as the dominant forces in the central place patterns shown in Figures 3.4 and 3.5. The "traffic principle" necessitates a revision of the system so as to ensure that as many places as possible lay on any one traffic route between two important towns, the route being established as straight and as economical as possible. The system then is essentially a linear one and might be expected to dominate, for example, in areas characterized by a ridge and valley terrain. By contrast the "administrative" or "separation principle" results in the creation of virtually complete districts of almost equal area and population, at the center of which lies the most important place.

Lösch's Theory

Christaller's main work was published in German in 1933. Seven years later there appeared an even more impressive book on the subject of the economics of location, also published in German, in which the theory of central places outlined by Christaller was developed further and generalized. Its author was August Lösch (1940).

Unlike Christaller, who began with an urban hierarchy to which functions were assigned, Lösch began at the fundamental level of a single economic activity, specifically a brewery, producing a good to be sold in a region. On the basis of his reasoning about the nature of the demand and the market area for this brewery, Lösch then sought to build up in logical sequence a system of market regions.

38

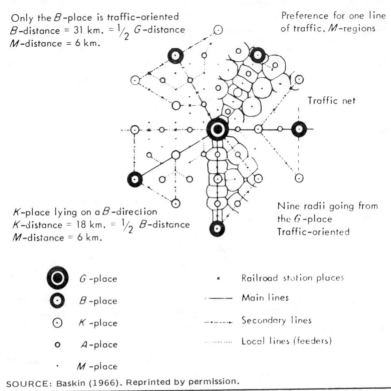

Only the *B*-place is traffic-oriented
B-distance = 31 km. = $\frac{1}{2}$ *G*-distance
M-distance = 6 km.

Preference for one line
of traffic. *M*-regions

Traffic net

K-place lying on a *B*-direction
K-distance = 18 km. = $\frac{1}{2}$ *B*-distance
M-distance = 6 km.

Nine radii going from
the *G*-place
Traffic-oriented

◉	*G*-place	×	Railroad station places
◉	*B*-place	·———	Main lines
⊙	*K*-place	—·—·—→	Secondary lines
○	*A*-place	·········	Local lines (feeders)
·	*M*-place		

SOURCE: Baskin (1966). Reprinted by permission.

Figure 3.4 A System of Central Places According to the Traffic Principle

THE SPATIAL DEMAND CURVE

The same initial assumptions about a homogeneous plain, uniform settlement distribution, complete transportation accessibility, and rational
economic behavior are made as in Christaller's discussion. Then, it is assumed that at some location in this plain a brewery is established. Bearing
in mind that the household must pay the transport cost from the brewery
to the farm, what will be the demand for beer by any one farm household
in the region?

In seeking to answer this question, we begin by considering the
household's general demand for beer. The diagram in Figure 3.6 suggests
that the higher the price for beer at the brewery, the lower the demand for

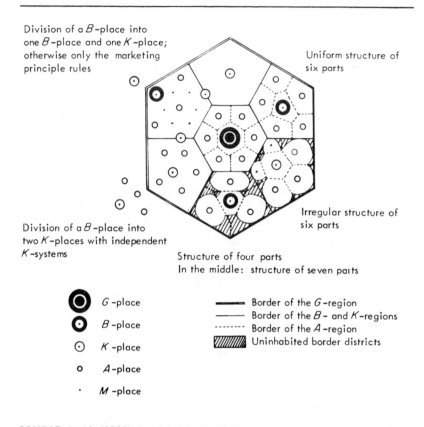

Division of a *B*-place into
one *B*-place and one *K*-place;
otherwise only the marketing
principle rules

Uniform structure of
six parts

Irregular structure of
six parts

Division of a *B*-place into
two *K*-places with independent
K-systems

Structure of four parts
In the middle: structure of seven parts

◉ G-place	━━ Border of the *G*-region
◉ B-place	── Border of the *B*- and *K*-regions
⊙ K-place	---- Border of the *A*-region
o A-place	▨ Uninhabited border districts
· M-place	

SOURCE: Baskin (1966). Reprinted by permission.

Figure 3.5 A System of Central Places According to the Administrative Principle

it. This is explained by the response of the household to the problem of deciding how to spend its income, which we assume is fixed, on beer and other things, given that the price of beer is rising. As the share of the income available for the purchase of the non-beer items falls, those items have to be sacrificed. But at some point, represented by the beer price p_0 in the diagram, this sacrifice becomes too great to tolerate and no beer at all is purchased. The particular shape of the curve in Figure 3.6 reflects the changing preference for beer as opposed to other items, over the range of possible beer purchases, and also the degree to which other items can be substituted for beer.

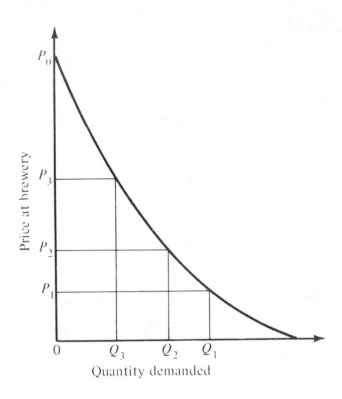

SOURCE: Based on Parr and Denike (1970)

Figure 3.6 Hypothetical Demand Curve for Beer

How is the demand for beer affected by the distance separating the farm household from the brewery? In Figure 3.6 three levels of demand—Q_1, Q_2, and Q_3—are identified corresponding to prices P_1, P_2, and P_3. Once the cost of transport is added in, the beer becomes more expensive the farther away the household is from the brewery; at some distance away the household cannot afford to buy it, and the demand is zero. These critical distances (S_1, S_2, and S_3), corresponding to the brewery prices (P_1, P_2, and P_3) and the associated levels of demand (Q_1, Q_2, and Q_3), are shown in Figure 3.7.

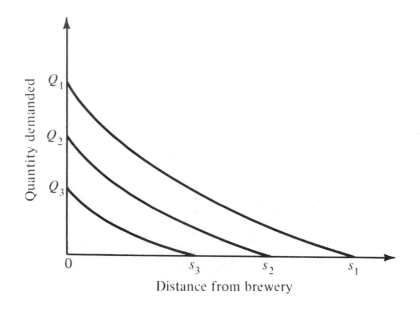

SOURCE: Based on Parr and Denike (1970)

Figure 3.7 Demand Curves as a Function of Distance from the Brewery

These relationships between price and demand can be summarized in a graph that shows the total demand curve facing the single producer of beer. This is the curve labelled D_1 in Figure 3.8. There is also shown in this diagram a curve (AC) that shows how the average cost of production for say a gallon of beer varies with the quantity produced. This curve shows that the average cost falls off, sharply at first, but then more gradually as the quantity is increased.

The price that is set initially by the one producer will be that at which the cost of producing each additional gallon of beer (this is referred to as the *marginal cost*) will be exactly equal to what the farmers as consumers are willing to pay for a gallon of beer. In Figure 3.8 assume that this price is P_1 and that the quantity of beer demanded (and, therefore, produced) at that price is Q_1. In the two preceding diagrams it was shown that this demand Q_1 was the demand at the brewery at that price but as transport costs were absorbed the demand fell off until it became zero at distance S_1 from the brewery. This distance is Christaller's *ideal range*. Each good

42

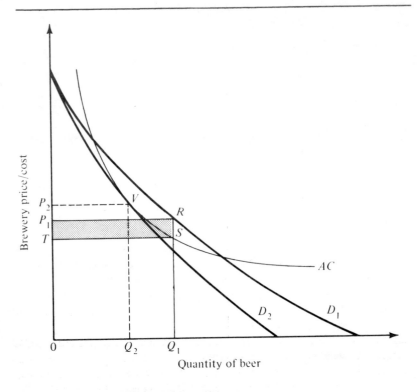

SOURCE: Based on Parr and Denike (1970)

Figure 3.8 Cost and Demand Curves for the Brewery

would have a different ideal range depending on the relationship that held for it between market price and transport costs.

The fact that in the situation described so far there is only one producer means that this one producer is obtaining *excess profits*. What are these? From Figure 3.8, it can be seen that at the level of output Q_1 the average cost curve AC lies below the demand curve D_1, which is also the average revenue curve for the producer. The average cost curve includes, it is assumed, some payment for the skills and labor of the producer himself, and, therefore, the difference between the points R and S on the graph, which is a monetary difference, is a profit in excess of this allowance. When it is computed over the total level of output Q_1, then the total excess profit is shown by the area of the shaded rectangle in Figure 3.8.

This attractive situation must appeal to other potential producers of beer who will be attracted into the business. As each new producer enters, a new central place will be chosen and the location pattern of breweries will adjust until each is serving an identically sized area and no excess profits are being realized. This means that for the original producer, there is less demand for the product at a particular price and the demand curve is shifted to the left (to D_2 in Figure 3.8). At some point this demand curve just touches the producer's average cost curve and at this point of tangency (V in Figure 3.8). with market price P_2 and output level Q_2, the equilibrium position is reached. If the producer seeks to raise his or her price above P_2, then demand will fall and average cost will exceed average revenue and the producer will go out of business. If the producer lowers the price below P_2 then excess profits will reappear, new firms will enter, and the price and quantity produced will be driven back to the equilibrium point.

It is important to note that every brewery will now have the same level of output because it is assumed that all firms in the same industry have the same cost structure and production schedules. Now that the price is P_2 then it can be seen from earlier diagrams that the demand at this price will fall to zero at distance S_2 from each brewery. This is the new ideal range. But with many producers now located on the plain, this ideal range will not be realized because at some distance away from one brewery the consumers will find it advantageous to buy from a different brewery. The boundary between the two market areas will set the limit on the real range of the good. This point was discussed earlier. Given the assumptions about the uniform distribution of costs and of farm households and so on, then the plain will be covered by a series of overlapping market areas for the breweries, and a pattern of hexagonal market areas will result with no area being unserved and no firm earning excess profits.

NETWORK OF MARKETS

The derivation of a hexagonal network of market areas is essentially a function of the space-filling requirement of central place theory. For example, when comparing the volumes of demand for a given good, Lösch argued that the demand in a cone with a hexagonal base is "2.4% greater than a square of equal size, about 10% greater than in a circle, if the empty corners are included, and at a maximum 12% greater than in an equilateral triangle of the same area" (Lösch, 1954, p. 113). As the superiority of the hexagon over the square is least, several researchers have suggested that urban systems based on either squares or hexagons could theoretically be constructed without undue loss of efficiency in location and service. In the strictest sense of obtaining the maximum packing density of places, however, the hexagonal distribution of places satisfies the criterion best.

In the earlier discussion of Christaller's work, it was noted that the number of market areas nested within higher order ones increased by a factor of three. This is the so-called k value and it refers to the number of equivalent full places of a particular order served by a place at the next higher order. For example, with reference to Figure 3.2 it was noted that the G-order place served the equivalent of three full B-order places.

Lösch contended that the k = 3 network of Christaller's scheme was but one special case of a more general arrangement, and that even the acknowledgment of Christaller's traffic principle (k = 4) and administrative principle (k = 7) did not exhaust the possibilities. Lösch considered as many as 10 of what he called these "smallest economic areas" for k values of 3, 4, 7, 9, 12, 13, 16, 19, 21, and 25.

Parr and Denike (1970) correctly observe that it is only in the very restricted sense of the geometrical appearance of the two schemes that Lösch's claim for the generality of his scheme over that of Christaller's holds true. A significant difference, for example, between the two formulations of Lösch and Christaller relates to the fact that in Christaller's scheme, as has been noted already, there is a *nested* hierarchy of places and market areas with the lower-order market areas completely contained within higher-order areas. Thus, a center offering good m, also offers all goods of a lower order than m. Lösch argued, however, that two centers of order m may have different functional mixes: That is, one may have a brewery and a bakery, while another may have a bakery and a laundry. Yet another center of similar size might conceivably contain all three functions. How could this situation arise?

The answer is provided in the way that Lösch generalized from his discussion of the market area arrangment for a single economic good to the regional level in which there are many goods offered by numeous producers in the urban centers of the region. The assumption of a uniform and continuously distributed population was dropped. The individual market area networks were then thought of as being overlaid on one another in such a way that there would be at least one common central location (the metropolis) and as many other place locations coinciding. Rotation of the networks around the central metropolis would then produce a landscape or location pattern in which there would be certain sectors with many urban places and others with few. These were referred to by Lösch as being "rich" or "poor"

45

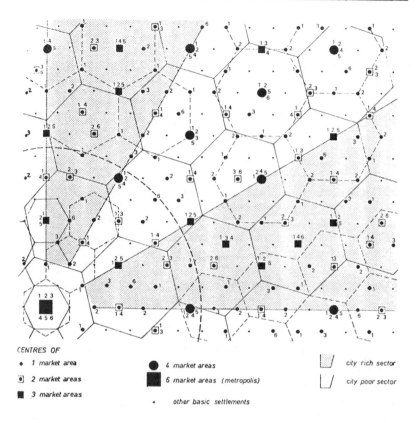

CENTRES OF

• 1 market area   ● 4 market areas   ⬒ city rich sector

⊙ 2 market areas   ■ 6 market areas (metropolis)   ⬓ city poor sector

■ 3 market areas   • other basic settlements

SOURCE: Saey (1973). Reprinted by permission from **Tijdschrift voor Econ. en Soc. Geografie** 64.

NOTE: "The market areas are ranked in ascending order of size. The entire lattice of market area 6 and parts of the other lattices are shown (below left: market area 1 and 2; below right: market area 3; above left: market area 4; above right: market area 5). The existence of city-poor and city-rich sectors is apparent from the location pattern in the part of the landscape within the dashed circle (the existence of the sectors in the other parts of the landscape cannot be demonstrated because of the small number of superposed lattices)" (Saey 1973).

Figure 3.9 Löschian Landscape Involving the Six Smallest Economic Areas

in cities and they are illustrated in Figure 3.9. In this diagram, the six smallest market area networks overlay one another.

Later writers on this topic have pointed out that the method used by Lösch in deriving his location pattern is not an entirely consistent one. They have suggested, for example, that Lösch underestimated the demand for different goods that would be associated with the population residing in the urban places themselves and that as a result excess profits would exist and other producers would be attracted into a center. But this would contradict Lösch's requirement that there be only one producer of a good in each place offering that good. These lines of economic argument flow naturally from Lösch's analysis, which is much more concerned with such issues than was Christaller's formulation.

Conclusion

The writings of the two scholars discussed in this chapter have stimulated a great deal of later research and scholarship in all fields of the social sciences. Numerous studies of the spacing of cities, the size and number of cities, the relationships between the population sizes and functional complexity of urban places, and the patterns of consumer shopping behavior have usually been prefaced with restatements of parts of central place theory and have then claimed to have tested certain of its predictions. In the next chapter some of these later contributions will be reviewed.

One other legacy of the work of Christaller and Lösch is that central place theory has been seen by many to provide a convenient framework for certain problems in regional planning. There are actual cases that can be pointed to in different parts of the world where settlement patterns have been planned in accordance with central place theory. Illustrations of this applied work are provided also in the next section.

4. SETTLEMENT PATTERNS, SOCIAL STRUCTURE, AND PLANNING

The theories outlined in the works of Christaller and Lösch, although published originally in German, had a dramatic impact upon the world of English scholarship.

Lösch's work was translated into English in 1954; Christaller's work was translated in 1957 by C. W. Baskin but was not published until some years later (Baskin 1966). Although their studies had been referred to in the English literature for more than a decade prior to these dates (a point that is elaborated upon in Harris 1978), nevertheless the timing of the publication of the full translations of both works could not have been more propitious for the fields of human geography and the newly emergent regional science. In both of these disciplines, and especially so in their North American schools, there were stirrings, more so in regional science than in geography, of interest in models, theories, and quantitative analysis. For the human geographers, long steeped in the tradition of urban analysis, Christaller's formulation appeared to offer an ideal framework for further studies of settlement patterns, urban functions, and trading areas. The findings of these studies could not only be related to the theory but inasmuch as they could be focussed on questions of urban size, numbers of economic functions, and central places and distance relations, they could be couched in the new language of quantitative analysis and be the object of inter-disciplinary attention (Müller-Wille 1978). This enthusiasm on the part of urban geographers was contagious and, as is illustrated below, numerous central place studies were completed for many regions around the world.

For much of this geographic analysis, Lösch's work was too economic and too theoretical in character to be of much interest. But for regional science with its greater emphasis on more formal theorizing and its bias toward economic analysis, Lösch's work was the preferred reference book.

The theories were not without their impact on other social sciences also. The interest in regional and urban analysis had been a slow one in terms of gaining momentum in economics but it is reasonably well established today. Lösch's work continues to provoke discussion in that setting.

More recently, in the fields of anthropology and archaeology central place theory, mainly of the Christaller version, has commanded considerable attention as a useful framework in analyzing social networks and structures both in ancient societies and presently underdeveloped ones. Examples of this work are provided later in this chapter.

The Geography of Central Place Systems

The geographers' interest in settlement patterns and in the role of urban centers has a very long tradition. Christaller himself acknowledged the contributions made by other geographers who were writing on the subject in the early decades of this century. More recently, Robic (1982) has suggested that a French scholar, J. Reynaud, outlined a theory of central places as early as 1841, predating by more than a half century those works that are generally acknowledged to have anticipated Christaller.

The overall stimulus given to geographic research by Christaller's contribution was a considerable one and it resulted in several new lines of work. These are reviewed here under three subheadings. First of all there were many studies made that sought to extend and duplicate Christaller's own empirical work on Southern Germany by analyzing the detailed trading relations of towns in selected regions, by identifying hierarchies among them, and by measuring their degrees of centrality.

A second and more difficult line of work sought to analyze the statistical relationships between selected features of central place systems with a view to confirming the generality of certain results. It is appropriate that the outstanding contributions of an American geographer, Brian Berry, be acknowledged in this context for it was he, more than anyone else, who developed this statistical analysis of central place systems to a very high level of expertise. Furthermore, he was able to show that the central place framework could be usefully applied in analyzing consumer trading patterns and hierarchical commercial structures within large metropolitan areas.

The third grouping of geographic studies is a much smaller set—that concerned with the use of the central place framework in the modeling and analysis of the diffusion of ideas and change across the surface of the earth.

HIERARCHIES AND CENTRALITY

Christaller's theory, as has been noted already, was strongly influenced by his knowledge of the urban places in the southern Germany of his time. Hence, when he turned to the task of testing his theory against reality, it was to that region and its towns that he looked.

The first challenge was to find some way of measuring the *centrality* of an urban place. For without such a means it would not be possible to identify the central places that belonged to the different levels of the hierarchy. The answer to the challenge, replied Christaller, was to rely on "a perplexingly simple and sufficiently exact method . . .: one need only count the telephone connections" (Baskin 1966, p.143). In accord with this suggestion, he proposed that the centrality of a place could be measured by a simple formula:

$$Z = T - P \ (TR/PR)$$

where Z was the centrality of a central place, T was the number of business telephone connections in that place, P was the population size of the central place, TR was the number of telephone connections in the region, and PR was the population of the region. In other words, the term T gave the actual observed number of telephone connections in the place while P (TR/PR) gave the expected number for the same place based on the assumption that the density of telephone connections per person was the same as in the region.

The number of telephone connections was regarded as the best available measure of the importance of a place. In Germany in the early 1930s private telephones presumably were rare and only business activities could boast of the service. Those public telephones that did exist were ignored by Christaller in his calculations.

The use of the index allowed Christaller to define more precisely the different levels of the central place hierarchy that he had assumed was present from the outset. That is to say, he did not calculate the index of telephone connections and then with respect to the values of this index identify the groupings of urban places. The seven main classes of centers shown now in Table 4.1 from the M up to the L places are those that were referred to earlier in the text (see Table 3.1). To each of these classes there was now attached a characteristic or typical value of the telephone index. The H places were not shown in the earlier Table 3.1. They were, in Christaller's view, only "auxiliary central places" and included a diverse group of small places, some of which were of declining importance while others were of increasing significance. Similarly, at the upper end of the hierarchy (although not shown in Table 4.1) were "the world cities or national capitals."

Christaller described the various levels of the central place system represented in Table 4.1 in considerable detail. There would be no point served in repeating those descriptions here. They did, however, convince Christaller that the marketing principle was "clearly dominant in determining the distribution of central places in southern Germany" (Baskin 1966, p. 192) and that explanations of deviations from the expected pattern could be found either in economic factors (the differential wealth of subregions or population density variations) or in noneconomic factors (such as historical forces or the physiography of the region).

Parr (1980) has pointed out that Christaller in his 1933 work claimed too much insofar as the supposed agreement between his model (column 2 of our Table 3.1) and the actual frequency distributions of central places by level was concerned. Parr noted that neither the $k = 3$ nor the $k = 4$ model yielded satisfactory descriptions of the real-world frequencies and that later attempts by Christaller and other writers such as Woldenberg (1968) to develop "hybrid" models of the $k = 3$, $k = 4$, and $k = 7$ structures proved just as unsatisfactory. Parr himself demonstrated that a "general hierarchical model" in which the value of k was allowed to vary from level to level within a hierarchy, did in fact yield predictions that were much closer to the observed frequencies of central places. The rationale he gave

50

TABLE 4.1 Centrality Values for Christaller's Southern Germany Towns

Type of Place	Population (approximate)	Number of Telephones	Centrality Index
H	800	5-10	−0.5-+0.5
M	1,200	10-20	0.5-2
A	2,000	20-50	2-4
K	4,000	50-150	4-12
B	10,000	150-500	12-30
G	30,000	500-2,500	30-150
P	100,000	2,500-25,000	150-1,200
L	500,000	25,000-60,000	1,200-3,000

SOURCE: Based on Baskin (1966, p. 158)

for the general model was couched in terms of three possible forms of structural change that might occur in a central place hierarchy. First, a new or intermediate level might be formed within a hierarchy either by way of centers being downgraded or upgraded in function. Second, the "extensiveness" of a level might change either through an increase or decrease in the number of centers at that level. Finally, a level of the hierarchy might disappear altogether.

To the modern reader, Christaller's applied work appears outdated, if only because the telephone has become such an integral feature of all private homes that the rationale underlying Christaller's index of centrality has disappeared. But it is worth noting that more than one later writer on the subject has used the numbers of long-distance telephone calls originating in small urban centers and directed to larger centers as indicators of the extent of those centers' zones of influence. J. D. Carroll, writing in 1955, reported on his analysis of the number of long-distance calls made from 50 small places in Michigan to the cities of Flint, Detroit, Lansing, Saginaw, and Bay City over a 10-day period in 1940. His results confirmed that the number of calls made to any one city tended to fall off as the distance away from the city increased, and that the larger the city the greater was its zone of influence. In a major study of urbanization in the upper midwest of the United States, Borchert and Adams (1963) also used the number of phone calls per capita from smaller centers to larger ones as a means of defining trade area boundaries.

A second feature of Christaller's applied work warrants a final comment. The classes or categories of centers in his urban hierarchy were established largely independent of the empirical work; the seven main levels shown in Tables 3.1 and 4.1 had been decided upon before any analysis of the information on telephone connections was completed. This feature of the work has been imitated in other studies that followed and it has been the subject of a great deal of criticism. One critic in particular, R. Vining (1955), argued that if a hierarchy of urban centers exists then its existence should be confirmed by the empirical analysis, not assumed beforehand. This particular issue is addressed in some of the statistical work that is reviewed later in this chapter.

Christaller's study spawned a large number of central place studies of many different areas all around the world. In their attempts to identify hierarchies of central places and to describe the sets of market areas associated with them, the researchers relied on typically one, or a combination of three different approaches. The first approach involved replicating components of Christaller's research by deriving an index of the level of centrality of each place based on some particular feature of the place; however, where Christaller had focussed on the number of telephone connections in a center, others, such as Godlund (1956) in Sweden, emphasized the number of shops located in the place and constructed an index based on this feature. Godlund used the following formula:

$$C = \sqrt{(B)\ (m)\ -\ (P)\ (k)}$$

where C was the centrality of the place, B was a count of the number of shops in the place, m was a weighting factor that took into account the comparative average size of the shops, P was the population of the town, and k was an index of the average accessibility to retail trade services enjoyed by any one inhabitant of the larger region in which the central place was located. In order to apply the formula, Godlund had to rely on estimates of m and k derived from other studies.

Once Godlund had computed the C values for all of the central places in his study areas in central and southern Sweden, then he combined these values with other information on the number of functions offered in the different places and produced a five-level ordering of the central places. He began with the observations that "places offering complete service—this taken to be that all the branches of trade listed in the inventory are represented—are practically identical with the places that have a C value of ≥ 5" (Godlund 1956, p. 36) The lower classes were then associated with "even one-degree intervals" from this top level. The resulting classification is shown in Table 4.2. The growth in commercial activity in the towns over the half-century was reflected in the increased values of the index.

52

TABLE 4.2 Godlund's Classification of Swedish Towns

Type of Place	C Values 1900-1920	1921-1937	1938-1951
D	4.0+	4.5+	5.0+
E	3.0-3.9	3.5-4.4	4.0-4.9
F	2.0-2.9	2.5-3.4	3.0-3.9
G	1.0-1.9	1.5-2.4	2.0-2.9
H	0-0.9	0-1.4	0-1.9

SOURCE: Based on Godlund (1956)

Godlund then turned his attention to the problem of defining the boundaries of the trade areas, or *umlands* as he called them, of the different classes of central places. By assuming that the extent or "intensity" of the field of influence of each urban place was related directly to its centrality and inversely to distance from the place, Godlund then was able to calculate the points (or locations) of equal intensity between each pair of places and to draw his boundaries accordingly. These forms of calculations are well known in market-area surveys where the interest lies in determining the lines of equal attraction, or equal market force, between pairs of centers. Godlund's results are illustrated in Figure 4.1.

A different approach to the problem of indentifying hierarchies and sets of market areas has involved the detailed analysis of the actual trading areas served by businesses located within the central place. The emphasis now is placed on determining the extent of the area served by the central place, and the indicators used are such things as the range of the circulation of a town newspaper or the extent of bus services provided by the town or, as has been noted earlier, the volume of telephone calls between different urban places. A good example of this approach is provided in Figure 4.2, which shows the hierarchical market areas for Estonia, now part of the USSR.

A third way of tackling this problem has been to conduct questionnaire surveys in the rural areas surrounding the central places to determine the levels of patronage and the trading areas of the centers. People are asked to identify the urban places where they go to shop for selected goods and services. This method is far from new and many of the rural sociological studies and the trading surveys of the early part of this century in the United States employed much the same strategy. The results of these types of surveys are typically represented in maps or charts showing for different

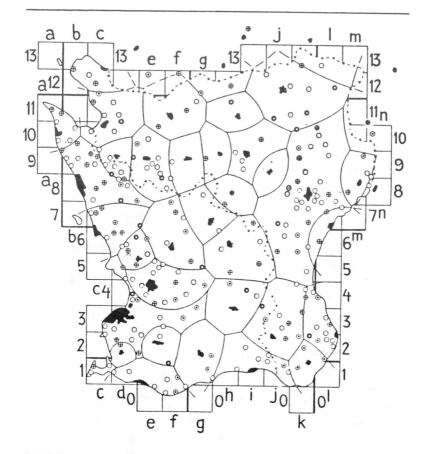

SOURCE: Godlund (1956). Reprinted by permission.
NOTE: The largest places were Mälmo (cell d3), Lund (34), Landskrona (c6), Hälsingborg (68), Kristianstad (k8), Ystad (il), and Trelleborg (el).

Figure 4.1 Theoretical Umlands of Godlund's D Places in Southern Sweden, 1950

goods and services the connections between the farmsteads surveyed and the urban places visited for the purpose of obtaining those goods and services. The results of such a survey are nicely illustrated in Figure 4.3. In this case, there was a particular interest in examining the cultural differences that existed in the patterns of shopping travel between "modern" Canadians and Old Order Mennonite farmers in an area of southwestern On-

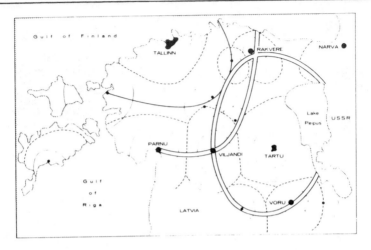

SOURCE: Based on Kant (1951), as adapted by Taaffe et al. (1970). Reprinted by permission.
NOTE: The lines are the boundaries for the tributary areas of the differed sized centers. Tartu and Tallinn are the highest-order places.

Figure 4.2 A Hierarchical Market Area System for Towns in Esthonia

tario adjacent to the "regional capital" of Kitchener-Waterloo; at the time of the study, 1963, this twin-city center had close to 100,000 people. The maps suggest that the more culture-bound and traditional Mennonites did more of their shopping for clothing and shoes at local urban centers than did "modern" Canadians who preferred to travel to the larger centers for these goods.

STATISTICAL ANALYSIS OF CENTRAL PLACE FEATURES

One of the approaches mentioned above to the problem of identifying hierarchies, namely that of focusng in on the economic activities located within the central place, obviously could be extended beyond the simple counting of telephone connections (as did Christaller) or shops (as did Godlund). For each central place, one could identify all of the functions offered by it and count the actual number of business establishments performing each function. In other words, one could determine how many banks, grocery stores, insurance agencies, pool halls, and so on were to be found in the urban place and compare these numbers with the corresponding ones for other central places. This is exactly the type of information

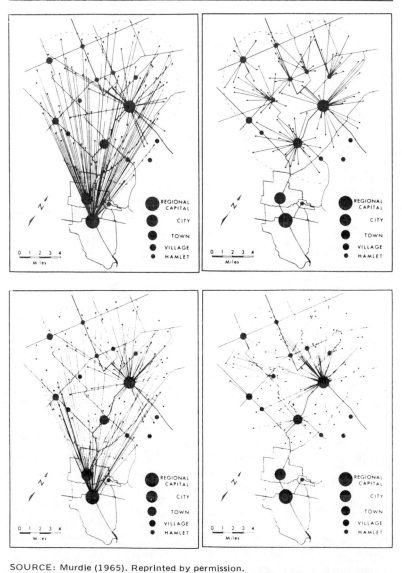

SOURCE: Murdie (1965). Reprinted by permission.
NOTE: The two right-hand diagrams show the travel patterns of the Old Order Mennonites in a Southern Ontario region for clothing (top right) and shoes (bottom right). The left-hand diagrams show the travel patterns of modern Canadians in the same region for the same purpose.

Figure 4.3 Consumer Travel Patterns for Different Cultural Groups

56

TABLE 4.3 Number of Establishments Offering Central Place Functions in Small Towns in Washington, United States

| | Central Places (population) | | | | | |
Function	Marysville (2,460)	Monroe (1,684)	Sultan (850)	Lowell (1,600)	Oso (200)	Threshold Population
Filling stations	9	8	6	1	2	196
Food stores	5	6	3	1	0	254
Churches	5	8	4	1	0	265
•	•	•	•	•	•	•
•	•	•	•	•	•	•
•	•	•	•	•	•	•
Physicians	3	3	1	0	0	380
Real estate agencies	6	2	0	0	0	384
Appliance stores	2	1	1	0	0	385
•	•	•	•	•	•	•
•	•	•	•	•	•	•
•	•	•	•	•	•	•
•	•	•	•	•	•	•
•	•	•	•	•	•	•
Department stores	1	1	0	0	0	1083
Optometrists	1	2	0	0	0	1140
Hospitals and clinics	1	0	0	0	0	1159
•	•	•	•	•	•	•
•	•	•	•	•	•	•
•	•	•	•	•	•	•

SOURCE: Based on Berry and Garrison (1958)

that is shown in Table 4.3 for a set of small urban places in Snohomish County in the state of Washington. In the study from which this example is taken, there were recorded as many as some 52 different central place functions for 33 urban places. The original table, however, is too large to reproduce here.

The study in question was an important one in the development of central place analysis. Its authors, B.J.L. Berry and W. L. Garrison (1958), proposed that these data could be used to estimate statistically the threshold populations of the functions and then to establish the hierarchies of both central places and functions. Their argument went as follows.

On the basis of the expected relationship between the population size of a central place and the number of functional units or establishments within it—a relationship that was referred to in Chapter 2 and illustrated in Figure 2.4—assume that the following equation holds:

$$P = A(B^N)$$

where P is the population size of a place, N is the number of establishments for a chosen function, and A and B are coefficients that have to be calculated from the information provided. There are well-established and conventional techniques for doing this form of calculation, known as "statistical estimation." These techniques are referred to as "regression analysis" and they may be studied in any number of excellent texts.

For each of the 52 functions illustrated in Table 4.3 and using the information on population size and numbers of establishments given in the table, the above equation was applied and regression estimates were obtained of the A and B values. Then the equation was reapplied using these A and B estimates together with a value of 1 for N to calculate an expected value for P. This was termed the threshold population for the function in question and it was obviously a surrogate measure of the threshold level of economic support that was referred to earlier in this text. It represented an estimate of the average level of population required to support one establishment of that function in the set of towns under study. These threshold values are shown in Table 4.3 in the right-hand column.

Once Berry and Garrison (1958) had derived the set of threshold values, then they ranked the functions according to these values and applied another form of statistical technique, in this case a grouping or classification procedure, in order to identify hierarchical levels. Three groups resulted. Use of the same procedure on a ranking of the urban places according to the total number of functions contained in them also established a threefold grouping of the central places. The two classifications were then merged together to define a hierarchical system.

The methodology first proposed in the Snohomish County study was later applied by Berry and his colleagues in an extensive study of central place systems in the American Midwest. This study has been the subject of numerous articles and monographs and was the basis of a book, *Geography of Market Centers and Retail Distribution* (Berry 1967). It is possible here only to summarize some of the study's main findings.

Four different areas within the Midwest were studied by Berry and his associates. Two of the areas were located in the state of South Dakota, the one a wheat-growing and cattle-raising area, and the other a cattle-raising, mining, and forestry area. Both of these areas were sparsely populated, at least in comparison to the third area, which was a uniformly settled section of the very fertile corn-growing and livestock-fattening region of southwestern Iowa. This third area was located in a state that, as noted by Berry et al., "has been the traditional, classic area for study of central places in the United States, because scholars thought that it satisfied the assumptions of central-place theory more nearly than any other region in North America" (Berry 1967, p. 4). The fourth and final study area chosen by the Berry group comprised three subareas within the city and suburbs of Chicago, Illinois.

The collection of information about the central place systems in these areas was conducted on two main levels. For the central places in the Iowa area especially, but also for the commercial shopping districts in Chicago, detailed information was obtained from questionnaire surveys of the type mentioned in the previous section of this review on the patterns of shopping trips made by rural consumers for particular goods and services. The results of these surveys served to demonstrate that hierarchies of urban places and of commercial districts did exist, that their generalized market areas could be delimited, and that the locations of the lower-order central places and commercial centers relative to the higher-order ones conformed to a regular pattern.

On the second level, statistical information was assembled from field surveys and the census on the population sizes of the central places, on the numbers of establishments performing central functions, on the areal extent of trading areas for selected establishments, and on the number of total persons served within these trading areas.

Once all of the information was assembled for the study areas, then it was possible to analyze it using both statistical techniques and graphs. For example, the statistical relationships between the different features mentioned above could be described using the regression techniques to which we have referred. We shall not spend time in reviewing these statistical results. Suffice it to say that all of the relationships that Berry examined were strong positive ones, which was reflected in the fact that the corresponding correlation coefficients were always above 0.87 and were positive in their signs.

The diagrammatic representations of the results of the studies are more readily comprehended than the statistical ones. For instance, Figure 4.4 shows all of the central places and urban commercial districts plotted as points according to the size in square miles of their trade areas and the level of total population served. The symbols V, T, and C in the upper part of the diagram represent the villages, towns, and cities of the farming areas in Iowa and South Dakota. Similarly, the symbols S, N, C, and R in the lower part and to the right of the diagram stand for the levels of the hierarchy of urban commercial centers, namely the convenience shops, the neighborhood centers, the community centers, and the regional shopping centers, respectively. Some general observations may be made about this diagram. First, in all areas there is a marked tendency for the total population served to increase as the size of the trade area increases (shown by the general lower left to upper right scatter of the points). Second, in regard to the scale in the upper right of the diagram (gross population density per square mile), which served to distinguish one study area from another, it is clear that as this overall population density increases (from upper left to lower right) trade area size decreases. This is hardly a surprising result in that if we recall the idea of a threshold level of population, it is to be expected that this threshold will be met within a smaller radius in a densely settled area than in a sparsely settled one. Finally, as Berry himself noted, "all lines separating levels slope backward to the left, indicating that trade areas increase in size as densities drop, but not as fast as the densities decline, so that the sizes of populations served fall" (1967, p.33).

An interesting application of the approach taken by Berry to the analysis of central place patterns was that by one of his students, H. G. Barnum (1966), in a study of a portion of the southern Germany region that Christaller had described. It was an exhaustive survey of more than 350 urban places (although only 244 were analyzed in detail), some 300 central place functions, and more than 1600 household interviews.

The statistical analyses performed by Barnum established a five-level hierarchical grouping of the urban places and served also to confirm the statistical relationships between the size and functional complexity measures that had been revealed in the U.S. studies. In the same vein, trade area surveys in the region pointed to the nesting of lower-order trade areas within the higher-order ones, again a result similar to that obtained elsewhere.

COMMERCIAL HIERARCHIES WITHIN METROPOLITAN AREAS

An additional comment or two should be made at this point on the subject of central place studies set within metropolitan areas. As was noted earlier, what Berry undertook in his analysis of the Chicago area was not entirely novel. Proudfoot (1937) had outlined a form of hierarchical classification of business centers into "local," "neighborhood," "regional," and "central" business districts, and Carol (1960) had used

60

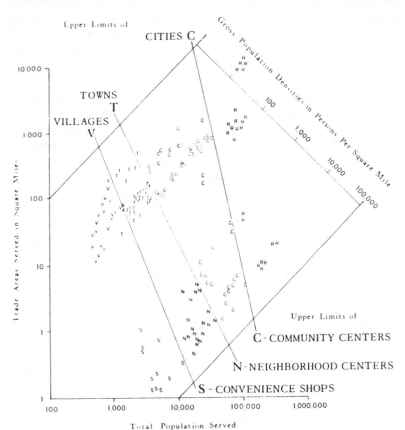

SOURCE: Brain J.L. Berry, **Geography of Market Centers and Retail Distribution**
© 1967, p. 33. Reprinted by permission of Prentice-Hall, Inc., Englewood Cliffs, N.J.

Figure 4.4   Relationships Between Size of Trade Area, Total Population
Served and Hierarchical Levels of Centers in Different Central
Place Systems

this same classification in his study of the city of Zurich. In that study,
Carol presented an even more fascinating illustration of the ranges of goods
offered at the different levels of the hierarchy. He noted, for instance, that
stores in a neighborhood business district generally stocked only one kind
of champagne, 112 kinds of cigars and wristwatches in the 40-200 francs
value range. At the higher-level regional business district, by contrast, there

would be available three kinds of champagne, 205 kinds of cigars, and 60-950 franc wristwatches. At the top of the hierarchy—in the central business district—as many as 20 kinds of champagne, 501 kinds of cigars, and 100-20,000 franc watches would be available (Carol 1960, p. 424).

Berry and his colleagues certainly extended the analysis of metropolitan commercial centers in a central place framework further than had been done before. A follow-up study of the whole of the city of Chicago, for example, focused on the hierarchical structure of its business centers as the framework within which to study the patterns, the probable causes, and the possible cures for economic and physical blight (Berry 1963). In Simmons (1964) the factors contributing to the dynamic character of metropolitan retailing districts were explored further and it was noted that

> there is a trend towards sharper differentiation of centers. The lowest order functions are disappearing or retreating from nucleations; the higher order functions seeking the higher order centers, are leaving the neighborhood centers . . . . The tendency for stores to seek higher level centers is a result of their larger scale and is part of the same process of increasing threshold which leads to the decline of scattered corner store businesses in the city and to the disappearance of the hamlet in the countryside. (Simmons 1964, p. 129)

Again, as a part of the overall research program Garner (1966) provided a detailed study of the locational preferences and arrangements of different businesses within the retailing centers of Chicago and showed how these were related to land values and to the comparative ability of firms to pay rent.

THE DIFFUSION OF INNOVATIONS

A different line of geographic inquiry that has drawn upon central place theory is that of analyzing the processes whereby new ideas or changes in practices are spread or diffused over the countryside.

The main stimulus for this interest in diffusion processes came from outside of urban geography. It was provided, in fact, by the work of Swedish geographer Torsten Hägerstrand. In a ground-breaking study published first in Swedish in 1953, Hägerstrand showed how the geographic spread of an innovation over time—specifically, the introduction of a new agricultural practice—could be analyzed and modeled with reference to such concepts as information flows, the decrease in interaction and contacts over distance, resistance to change, and barriers to the flow of information and ideas. This study and the subsequent discussions of it in the geographic literature were to have almost as great an impact upon the development of quantitative and theoretical human geography in the 1960s and later as had Christaller's work.

Numerous studies of the spatial spread of new ideas and practices—for example, the introduction of television into communities, the adoption of tractors into farming areas, the use of new irrigation practices, or varieties in flows of information—were cast in the mold of the Hägerstrand model.

One suggestion made by Hägerstrand in a 1966 publication (p. 27) was that in the diffusion of ideas and innovations over the surface of the earth,

> the spread along the initial "frontier" is led through the urban heirarchy. The point of introduction in a new country is its primate city; sometimes some other metropolis. Then centers next in rank follow.

This idea that there is a diffusion down through urban hierarchy story was incorporated subsequently into a number of diffusion studies. Berry (1972, p. 118) for example, in discussing the diffusion of television stations throughout the United States from 1940 to 1968, noted that "the time-path was essentially hierarchical; the smaller the city, the later the opening of its TV station." Huang and Gould (1974) drew a similar conclusion in regard to the diffusion of branches of the Rotary Club throughout the United States from 1905 to 1972.

In an earlier study of the spread of the disease cholera, throughout the eastern portions of the United States in the nineteenth century, Pyle (1969) had found evidence of a hierarchical process (Figures 4.5 and 4.6). The disease first appeared in New York in May of 1866 and then spread down the eastern seaboard and also over into the Ohio-Mississippi River valley towns. Although there were exceptions to the rule, generally the larger cities were hit first and as the epidemic progressed smaller and smaller communities reported occurrences of the disease.

The interaction in the form of trade and social contacts between rural areas and urban centers and between urban centers of different orders, that we have highlighted in our discussion of central place theory, suggests a mechanism whereby new ideas and innovations may be spread through the hierarchy. Associated with the trading and commerce are exchanges and flows of information that provide the medium for the spread of new ideas, perhaps about new products, practices, or attitudes. These flows of information may be thought of as being channelled from one center to another depending on their relative standings in the urban hierarchy and, hence, on the dominance relations that govern them. Recall that higher-order places dominate lower-order ones, and in this formal central place hierarchy the

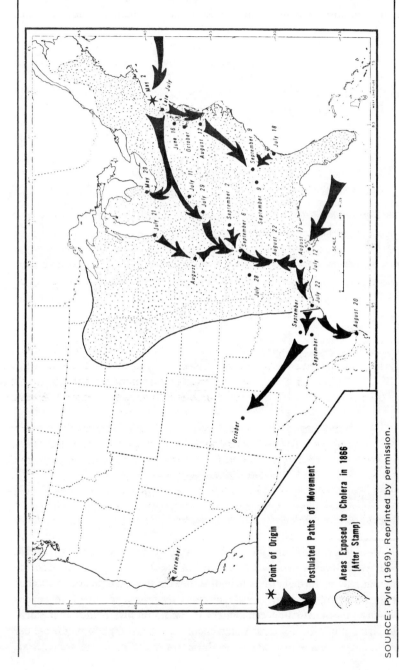

SOURCE: Pyle (1969). Reprinted by permission.

Figure 4.5 The Geographic Spread of Cholera in the United States, 1866

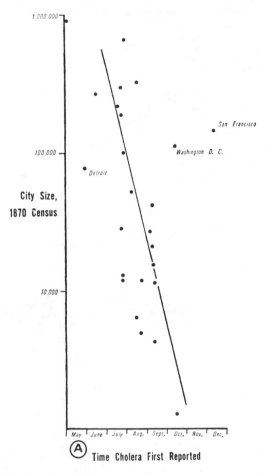

SOURCE: Pyle (1969)

Figure 4.6 Occurrence of Cholera in Different-Sized U.S. Cities, 1866

flows of information and the associated spread of innovations will always be from the higher-order center to the lower-order one—never in reverse. In fact, if the formal properties of a central place hierarchy are stated mathematically and if further assumptions are made about the levels of interaction that occur between centers in the hierarchy, then it is possible to derive other very formal statements, again expressed in mathematics, about the process of diffusion down through a central place hierarchy. Hudson (1972) developed this line of argument. He showed how mathematical equations could be derived and solved to describe the diffusion process in terms of the number and proportions of centers of particular orders that had been reached by the process after stated numbers of time intervals.

Hudson also discussed two other forms of hierarchical diffusion that differed in structure from the central place pattern of diffusion in which every center dominated specified sets of centers of lower orders. In the first of these two options, it was assumed that any center of a particular order dominated only a set of centers of the next lower order and no others. In this case, "diffusion takes place strictly according to size, proceeding from largest to smallest" (Hudson 1972, p. 122). In the second option, it was assumed that diffusion took place randomly within the hierarchy and that no specified arrangements of the centers according to either size or location had any controlling influence on the process.

For both of these options, Hudson showed how it was possible to derive expressions, corresponding to those obtained for the central place system, for the numbers and proportions of affected or contacted centers at some time t.

An interesting application of the formal models developed by Hudson was provided by Weisbrod (1974). He attempted to describe the spread of wage inflation down through an urban hierarchical diffusion process. The different levels of the hierarchy were defined in terms of what he called "bundles of occupational skills," with the highest-order centers possessing the most specialized skills. At the top of the hierarchy, in the highest occupational skill bundle, the demand for that labor was assumed to be boosted by some external force (new markets for the products, for example) to the point where supply was insufficient to meet demand. This excess demand for the occupational skills in question generated an inflationary impulse, or wage boost, which spread down through the other occupational skill bundles and, eventually, the other levels of the hierarchy. Unfortunately, the data that Weisbrod could obtain did not allow for the detailed testing of his model, and he was able only to demonstrate that certain patterns of dominance of large centers over smaller ones held in regard to the timing of wage increases.

The process studied by Weisbrod—that of explaining how wage increases spread throughout a large region—illustrates the limitations of purely hierarchical models of diffusion, especially in situations where the process under review may not be strictly one of diffusion. Wage settlements, it is known, are subject to a number of influences and forces. National unions often are involved and they may press for wage increases across the country. Also, the organizational structure of the industry involved may have more influence on the wage settlement process than does its locational pattern. If the industry is organized into several branch plants with each dependent on the other, then this arrangement may have more influence on the wage determination process than will the fact that the branch plants are all located in communities of roughly the same size and, hence, at the same level in the urban hierarchy.

It is reasonable to conclude that for those processes that do seem to spread over geographic space and through time in more or less discrete steps, a mixed mode of explanation is probably required. There may be within the process elements of a wave-like spread together with both hierarchical and neighborhood effects. It certainly appears to be the case that very few, if any, phenomena are spread by way of a strict central place hierarchical process. Pred (1977) has stressed this point in regard to the circulation of information in early nineteenth century urban America.

Anthropological Studies of Social Structures

Central place theory, as has been mentioned already, has an even wider appeal than that which it has enjoyed among geographers alone. Cultural anthropologists and archaeologists, in particular, have found that the theory provides a useful framework for analyzing the social structures of societies that either have existed and flourished in the past or exist today in underdeveloped countries.

The studies by G. W. Skinner (1964, 1965, 1976a, 1976b) of Chinese society from the thirteenth century up to the present provide the best examples of this line of scholarship.

The traditional agrarian society of China in the countries preceding the formation of the People's Republic in 1949 was, according to Skinner (1964, p. 43), "at once derivative of and enmeshed in two quite distinctive hierarchical systems—that of administration to be sure, but that of marketing as well." The first of these two hierarchies was more easily identified, consisting as it did of three main levels below the imperial capital Bejing. There were the provincial capitals, the prefecture capitals, and the county or district capitals.

The description of the marketing hierarchy provided by Skinner was couched in the language of central place theory. At the lowest level of the hierarchy was what he referred to as the "standard market town." This "provided for the exchange of goods produced within the market's dependent area, but more importantly it was the starting point for the upward flow of agricultural products and craft items into higher reaches of the marketing system, and also the termination of the downward flow of imported items destined for peasant consumption'' (Skinner 1964, p. 6). Each of these standard market towns would typically be dominated by two or three higher order centers, the arrangements of which were regarded as suggestive of Christaller's $k = 4$ and $k = 3$ networks, respectively.

The complete marketing hierarchy involved as many as seven or eight different levels that were differentiated from one another by the number of retailing activities present, the complexity of the services for offering credit and supplying retailers, and the presence or absence of guilds and business associations. It is interesting to note, however, that in the actual task of classifying central places into the levels of a hierarchy, Skinner emphasized the level of postal service available in the center. This was his indicator of the level of importance or centrality, and it brings to mind Christaller's telephone index and Godlund's shop-counting process. Skinner (1976b, p. 348) noted that

central places were classified according to available postal facilities as follows: (1) first-class PO; (2) second- class PO offering three or more special services including express delivery and money orders up to 100 yuan; (3) second-class PO offering two or more special services including express delivery and money orders up to at least 50 yuan; (4) second-class PO offering two or more special services that did not include express delivery; (5) second-class PO whose only special service was the availability of money orders not exceeding 50 yuan; (6) second-class PO offering no special services; (7) third-class PO; (8) agency only; (9) no postal facilities of any kind.

The manner in which the marketing hierarchy interlocked with the administrative one is illustrated in Table 4.4. What is suggested by this table is that very, very few of the lower-order marketing central places—that is, of the standard and intermediate market towns—served also as administrative centers, but that the degree of overlap between the two hierarchies increased at the higher levels. Even then, there were many higher-order centers in the economic hierarchy—for example, regional cities that served large trading areas but had only minor administrative roles as county centers. The administrative hierarchy was interpreted as an expression of the political forces within the country while the economic hierarchy reflected those of trading relations and commercial power.

A major question posed by Skinner is one that has not yet been addressed in this book and that has to do with the growth and development of a central place system over time. How does it occur? Christaller, admittedly, did consider the question of change in the urban pattern that he described,

TABLE 4.4 Distribution of Central Places by Level in the Administrative and Economic Hierarchies[a]

Level in Economic Hierarchy	Level in Administrative Hierarchy					Total
	Imperial	Provincial[b]	Prefectural	County	Nonadministrative	
Central metropolis	1	3	1	1		6
Regional metropolis		15	4	1		20
Regional city		1	41	13	8	63
Greater city			95	86	19	200
Local city			69	498	102	669
Central market town			17	580	1,722	2,319
Intermediate market town				106	7,905	8,011
Standard market town				12	27,700	27,712
Total	1	19	227	1,297	37,456	39,000

SOURCE: Skinner (1976a)

a. Agrarian China, except Manchuria and Taiwan, 1893.
b. Includes Nanking, "secondary" imperial capital and seat of governor-generalship of Anhwei, Kiangsi, and Kiangsu.

as Preston (1983) has reminded us, but his generalizations on this subject were far less imaginative ones than those that he presented for the static pattern.

In his studies, Skinner, of course, was presented with a unique opportunity to consider this question of evolution. The Chinese urban pattern is one of the oldest surviving ones in the inhabited world; and where most writers might seek to describe changes over a decade or two, Skinner could point to changes over a period of time of some six centuries! He observed, for example,

> In A.D. 1227, during the Southern Sung, the four hsien on the peninsula in Chekiang . . . supported twenty-six rural markets; six and a half centuries later . . . there were approximately 170 rural markets in the same territory. (1965, p. 195)

The process of the growth of the central place hierarchy was seen as resulting mainly from the addition of new households, villages, and markets. The number of "deaths"—that is, settlements that went out of existence—Skinner ignored as being inconsequential. The growth process was described in six main stages. At the outset, each dispersed market town was surrounded by a ring of six villages. As the landscape developed this inner ring was girdled by a larger one of another twelve villages. Beyond this second step, the process was depicted as a two-track one. For the $k = 4$ network, new settlements appeared mainly on the existing routes between old villages and markets, whereas in the $K = 3$ network they were established at locations equidistant from three existing centers. The intensification of the process and the infilling of the region with settlements resulted in two final forms, each with a strong hierarchical pattern of market towns established. In the mature $k = 3$ network there were 18 villages served by the market town and this has prompted some writers to describe Skinner's model as a $k = 19$ one. This is so only at the lowest level, however; above that one it is the $k = 4$ and $k = 3$ ratios that hold.

The $k = 3$ model, Skinnner believed, was favored in China not only because much of the urban development was in river valleys and basins that more closely approximated the uniformly fertile and accessible plain assumed by Christaller but also because of the proximity of large cities. These, Skinner observed, not only provided a "lavish production of nightsoil" that was available to improve the fertility of the surrounding farmland but they also tended to have better road systems developed around them. By contrast, the $k = 4$ networks were more typical of those areas that were "hilly to mountainous in topography, relatively deficient in arable land, and removed from large urban centers" (Skinner 1965, p. 204).

Skinner's approach toward analyzing the social structure of an agrarian society through a very detailed examination of the organization of its marketing patterns and the hierarchy of urban centers that are the focal points of the trade and commerce, has been adopted subsequently in many anthropological and archaeological studies. A number of such contributions were collected together in a two-volume work on regional analysis edited by the anthropologist, Carol Smith (1976a, 1976b). In introducing the first volume, Smith observed the following:

> One notable feature of the classical central-place is that they seem to be more often found in agrarian societies, where central places are periodic and traders mobile." (1976a, p. 15)

It was conceded that this was not surprising in view of the facts that such societies had dispersed populations and that mobile traders could respond much more easily and rationally to patterns of economic demands than could fixed markets. However, for those regions where permanent urban settlements did exist and trading arrangements were fixed, Smith believed that it would be necessary to propose new models or modifications of the existing ones. In this vein, a number of the papers in Smith's first volume discussed variations on Christaller's central place model. An example is shown in Figure 4.7, which is a composite model of the $k = 3$, $k = 4$, and $k = 7$ networks derived on the basis of a study of western Guatemala.

These descriptions and generalizations of the central place systems, based largely on field observations, were seen as yielding "typologies of regional systems that can ultimately explain a great deal about the social as well as the economic order" (Smith 1976b, p. 7). With this goal in mind, the authors in the second volume of Smith's collection sought to elaborate on particular patterns of social organization against the background of the spatial and hierarchical organization of the societies in question. Crissman (1976) in a study of marriage patterns in western Taiwan, for example, found that "marketing has a causative effect on marriages" (p. 147) in the sense that marriage partners were more likely to come from urban centers that had higher levels of marketing interaction. Similarly, Adams and Kasakoff's studies of "endogamy" (the tendency of people to marry within their own social group) found that the use of central place theory "demonstrated the existence of systems of interlocking groups above the village level, groups that are significant for marriages as well as for marketing" (1976, p. 186). They added a disclaimer, however, to the effect that there were interactions that did not appear correlated with marketing structures.

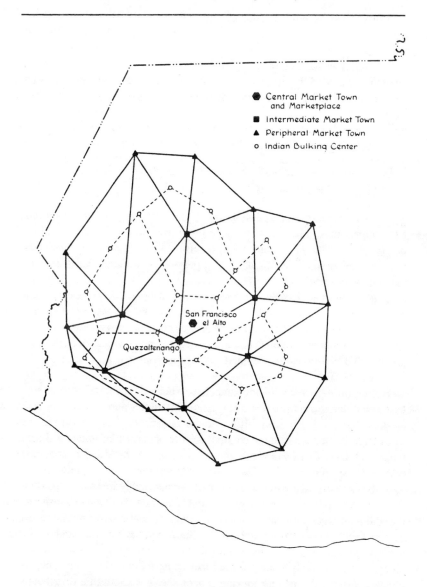

San Francisco
el Alto

Quezaltenango

● Central Market Town
   and Marketplace
■ Intermediate Market Town
▲ Peripheral Market Town
o Indian Bulking Center

SOURCE: Smith (1976b). Reprinted by permission.

Figure 4.7   Central Places in Western Guatemala

72

The final word in this section on the analysis of social structures is left with Skinner (1976b). He noted that his studies of China had underscored the "degree to which cities were at once embedded in society and essential to its overall structure" (p. 346). The structure of the urban system, the number and size of the urban centers, and the coincidence and overlapping of their market areas "determined variation in urban social structure" (p. 346).

Central Places and Regional Planning

One of the other legacies of the work of Christaller and Lösch is the fashioning of regional development plans around the postulates of central place theory.

The reference to central place theory in such planning endeavors typically has implied an acceptance of the idea that a well-developed, hierarchical central place system is in some sense an efficient arrangement that is likely to have a positive or beneficial effect upon the economic development of the region in question.

This assumption has been made not only in regard to developing but also developed economies. In the case of the former, Johnson (1970), for example, suggested, "The neglect of central-place analysis in the planning techniques of underdeveloped countries is doubly unfortunate: opportunities are lost and resources are devoted to less than optimal uses, perhaps far less" (p. 137).

In the same spirit, Mabogunje in his 1980 book on the development process noted that the central place models of Christaller and Lösch "do have an underlying value premise of economic rationality, and some measure of distributive justice in their concern for the provision of goods and services to all individuals in a given society" (p. 202). In regard to a developed economy, Simmons (1975) has presented a similar line of argument.

There are specific planning efforts that can be cited in illustration of what is argued for by Johnson and Simmons. In the case of developing economies, the planning study of Ghana prepared by Grove and Huszar (1964) is an example of a study that devised a hierarchical pattern of urban centers for the country (Figure 4.8), and proposed the upgrading of existing places along with the establishment of a number of new urban places in order to guarantee that "the increasing quantity of services should be distributed between centres to ensure the maximum benefit from the resources available" (Grove and Huszar 1964, p. 66). Their plan, however, was not adopted in the form proposed.

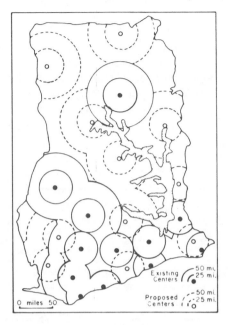

SOURCE: Based on Grove and Huszar (1964), as adapted by Taaffe et al. (1970).
Reprinted by permission.

Figure 4.8 Proposed Arrangement of Urban Centers in Ghana

A different approach to the planning of a hierarchical urban system in a developing country was taken by Harvey et al. (1974). They began with the existing location pattern in Sierra Leone and sought to devise a mathematical solution to the problem of determining where urban centers of different orders in a hierarchy should be located (or encouraged to develop) in order to provide the best possible patterns of access for the population to the centers offering different goods and services. The mathematical solution simply ensured that no cluster of population was more than a certain prescribed distance away from a central place. Again, there is no evidence to suggest that this plan was ever adopted.

One approach to central planning in a new economy that was put into practice involved the state of Israel. Following the creation of that nation in 1948 its newly formed National Planning Department devised a plan for development that was aimed at achieving uniform regional growth in part through the establishment of Christaller-type urban hierarchies in each region

(Brutzkus 1974; Schachar 1971). The plan called for five main levels of settlements—the village unit of 500 people; the rural center serving four to six villages; urban places of 6,000-12,000 serving tens of villages; towns of 15,000-60,000 population serving as regional centers; and national centers each of more than 100,000 population. These centers were established in most regions but as a number of later commentators have observed (Berry 1973; Mabogunje 1980), the plan was only moderately successful in that the emergence of a strong cooperative *kibbutz* system tended to work against the development of strong rural to urban interactions and favored instead the strengthening of direct relations between the kibbutzim and the larger urban centers.

Applications of central place theory in the planning of more developed and mature economies have been just as numerous. The locations of new settlements on the drained polders of the Netherlands were planned in this manner, for example (Thijsse 1962). In Canada, the Saskatchewan Royal Commission on Agriculture and Rural Life (1957) presented a detailed analysis of the central place systems existing in that Province (Figure 4.9) and urged that "public policy encourage the reorganization of rural services on the basis of the service center principles demonstrated in this report" (p. 139). The focus of attention for such public policy was recommended to be in the reorganization of local government boundaries and in the redefining of school district boundaries so as to ensure closer conformity with the existing trading areas of the urban centers.

The above proposals were never acted upon in any direct manner. This was not the fate of similar proposals which were made in Sweden. Pred (1973, pp. 5-9) has reviewed the impressive number of planning actions in that country that were firmly rooted in central-place-type studies. The locations of new schools and the boundaries for school districts, the location patterns for community centers or assembly halls, and the definition of new municipal administrative units were all accomplished using the principles of central place theory. In the case of the problem of defining the new municipal boundaries, Godlund's method for determining the zones of influence or umlands (which was discussed earlier) was the basis of the recommendations.

All of these efforts aimed at incorporating central place concepts into regional planning designs should be viewed in the total context of regional planning. In that broader setting they do not appear particularly impressive if only because of the fact that so much effort in so many parts of the world has gone into other approaches to regional planning and development. These efforts have been targeted mainly at the existing patterns of regional

75

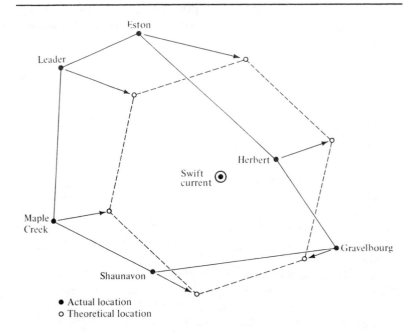

SOURCE: Based on Saskatchewan Royal Commission on Agriculture and Rural Life (1957)

Figure 4.9 Arrangement of Central Places in Saskatchewan

economic inequality and have sought by way of the use of different economic mechanisms (tax incentives, development grants, manpower training programs, public works, and so on) to stimulate growth in lagging regions and to remove the barriers to the flows of the factors of production among regions. In these approaches the pattern of urban settlements is taken as given. Where urban development has been the focus of attention, greater emphasis seems to have been given to the idea of growth centers and to the notion that investment and development in few such key places would have a stimulating effect upon the economies of the regions in which these centers are located. Rather than call for an even and balanced pattern of growth such as might be expected in a well-structured central place system, these plans were predicated on the idea that unbalanced regional growth held the key to success. These urban development strategies have often been referred to as "growth-pole" plans.

The conclusion drawn here is not that central place notions are completely irrelevant in regional planning and development efforts. Rather, it

76

is one that they have much to contribute to an understanding of how regional economies and social systems function and should be, therefore, an integral and important component of any regional development plan. Central place theory, we have seen, is much more than the simple hexagonal pattern of trading areas that is portrayed in neat diagrams. It is a theory about processes—economic, social, and behavioral—and these are ignored at the planner's peril.

5. NEW STATEMENTS OF THE THEORY

There are innumerable books and articles in philosophy and related disciplines on the subjects of what constitutes a theory and how any theory is proven correct or incorrect. These philosophical arguments are not of particular interest to us here, but one or two of them are hinted at in the following account of what has happened in the recent development of central place theory.

Recall that this theory, as developed by Christaller and Lösch, consisted of a set of statements that were logically connected and logically consistent, derived within the framework of certain assumptions. Webber (1971) suggested that these assumptions consisted of two subsets. The one contained four "environmental" assumptions, namely, "(i) there is a static economy upon (ii) an unbounded, homogeneous plain in which (iii) changes in agricultural location do not affect town location and in which (iv) we consider only industries and firms which sell goods to town hinterlands" (p. 17). These described the setting of the theory. Within it, both producers and consumers were bound by six "behavioral" assumptions which were "(i) firms choose maximum profit locations, (ii) all points are served by some firm in every industry, (iii) no firm makes abnormal profits, (iv) areas of supply and sales are as small as possible, (v) consumers patronize the nearest seller and that (vi) society maximizes the degree to which firms agglomerate (that is, minimizes the number of place sites)" (p. 17). Saey (1973) discussed these behavioral assumptions in a more terse style. He suggested that they could be summarized in two fundamental statements: Specifically, that consumers behave as if they all travel to the nearest center, and the entry of producers or suppliers into the market is organized such that given the environmental assumptions, the number of centers is as small as possible.

In the articles by Webber and Saey there is illustrated nicely the sharp difference of opinion about the testing of social science theories that often surfaces in such discussions. On the one hand, Webber suggests that verification of the theory cannot proceed by way of determining whether the behavioral assumptions hold in reality; for, in fact, the behavioral assumptions are unobservable. One cannot tell, he contends whether firms have chosen maximum profit locations or whether abnormal profits are absent and the maximum number of firms is present. The alternative approach to verification—that is, to find some area "where some environmental conditions are identical to those assumed" and "to observe whether the predictions of central place theory actually obtain there" (p. 23)—is considered impossible even for "the near-homogenous state of Iowa." There are simply too many other forces at work in such an area that cannot be held constant or assumed away." Webber's conclusion is that "the predictions of central place theory have never been empirically tested and that the theory has been neither confirmed nor disconfirmed" (p. 27). Buursink (1981) added his support to this contention.

Saey (1973, 1980, 1981) in his writings on central place theory adopted a more positive attitude in regard to whether the theory can be tested or not. While conceding that certain observable facts did appear to contradict the theory, nevertheless, he argued, this should not prompt abandonment of the theory but should rather lead to further research in an effort to refine it. Nor did Saey agree with critics such as Buursink (1981) who insist that the behavioral predictions of central place theory cannot be tested. On the contrary, Saey and Lietaer (1980) showed how operational tests could be made of the hypothesis that consumers behave as if they all travel to the nearest center, and they presented results in support of the hypothesis.

There have been other attempts made to refashion the central place theory by changing the assumptions, in particular the behavioral assumptions, on which it is based. For example, Rushton (1971) described a central place pattern derived on the basis of a modified assumption about consumer behavior that is summarized in the following section ("Consumer Behavior and Central Places") of this chapter.

The assumptions and conclusions about the economic foundations of central place theory—the arguments about the minimum costs and profit levels—have also been reexamined in the works of the economists Tinbergen (1968), Bos (1965), and Gunnarsson (1977). Their formulations of a system of production centers are outlined in the third section of this chapter, "Dynamics of Central Place Systems." Only an outline is given because the studies involved are quite mathematical in character and rely on techniques and proofs that are too advanced for this introductory statement.

The same is true of the work of White (1977, 1978) who has sought to relax the assumption of a static economy in the central place theory and to deal with a dynamic system of places. He describes the change using sets of equations, and so again it is possible here only to outline his main arguments.

The interest in writing the central place theory in more formal mathematical language, and in generalizing it in the process, reached a high point in two recently published monographs that are briefly summarized in the next to last section of this chapter, "Formal Restatements of Central Place Theory."

"Alternate Approaches to Settlement Theory," the concluding section, refers to some quite different approaches to the study of central place systems. These contributions to settlement theory discard most of the assumptions of the Christaller-Lösch formulation and work within much more simplified frameworks that require only one or two behavioral assumptions.

Consumer Behavior and Central Places

In the classic central place model, as has been noted at the start of this chapter, it is assumed that each rural household will patronize the closest central place that offers the goods and services demanded by the household. But the empirical studies of rural shopping patterns, such as were illustrated earlier, have confirmed that there is typically some variation in the choices made by the rural population as to where they actually go to shop. In many cases they may bypass a nearby center in preference for a more distant and usually larger place. There are a number of obvious explanations for such behavior. The larger centers may provide a wider range of choice in the goods and services sought and also offer the opportunity, because of their greater functional complexity, of satisfying several needs or purposes with one shopping trip. A visit to the city obviously affords one an opportunity not only to do one's banking and grocery shopping but also to take in a show or to visit an art gallery! This multiple-purpose shopping is far more restricted in the smaller centers. Further improvements in transportation have increased considerably the distances over which people are prepared to travel on their shopping trips and the closest center may not be the most attractive destination.

These observations about the patterns of rural shopping and trip making prompted Rushton (1971) to suggest that it was more appropriate to describe this behavior in terms of the preferences that consumers have for shopping in different centers and to then examine the effect of this proposal on the structure of a central place system. As an aside at this point, before continuing with the discussion of Rushton's models it is worth noting that Saey and Lietaer (1980) have argued that the central postulate of Christaller's theory—that consumers patronize the nearest center—should be regarded as allowing for some individual deviations and that it is, after all, the aggregate effect of consumer behavior that really matters. Their own empirical studies proceeded on that understanding and they sought to determine whether the frequencies of trips made to different Belgian centers were consistent with the Christaller postulate, given that these centers varied in both their sizes and locations from the hierarchical urban network assumed by Christaller. They concluded that they were supportive of the postulate.

Rushton assumed that each rural customer goes through a decision-making process during which *preference structures* for the different urban places are formed. These preference structures may be influenced by variables such as the number of functions offered in a center, the size of the place, and its distance from the consumer's home. The resulting patterns of shopping trips made to the places and the development of these centers in turn depend on their relative attractiveness, as reflected in the preferences of the consumers.

In seeking to demonstrate these effects, Rushton used the hypothetical set of indifference curves depicted on Figure 5.1. These show a series of trade-offs between size of town and distance from potential consumers. For example, the rightmost curve indicates those sets of distance-town size combinations where 10 percent of the population would make particular trade-offs. That is to say, 10 percent of the population would show this indifference between the size of town visited and the distance that had to be traveled. They would be indifferent between traveling say 28 miles to a town of around 10,000 population or traveling about 14 miles to a town of 2,000 population. By contrast, the left-hand curve shows the distance trade-off that would be acceptable to 90 percent of the population, and it suggests that the majority of persons are much less willing to travel longer distances to particular-sized towns. This 90 percent, for example, would prefer to travel only about 10 miles to towns of 10,000 or so population.

By using a computer simulation model, Rushton was able to examine the impact of these preference structures on the size distribution of towns in a hypothetical central place system. Not surprisingly, some centers failed to achieve the size suggested for them in a Christaller-type model while other places showed up as being larger.

In a recently published paper, Forster and Brummell (1984) have examined further the question of what happens to a central place system in which multiple-purpose shopping trips occur. They find that the tendency for producers to concentrate in the larger centers is increased; that con-

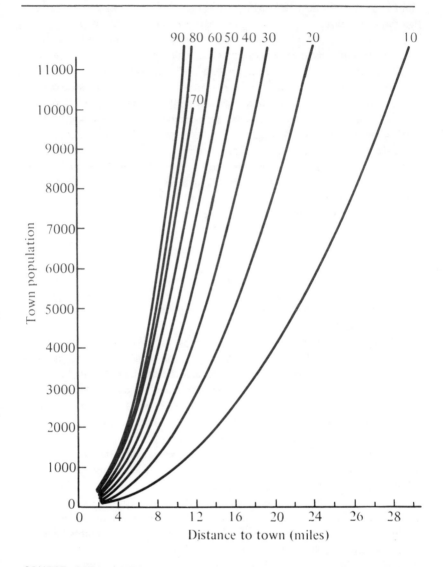

SOURCE: Rushton (1971). Reprinted by permission.

Figure 5.1 Preference Curves Showing Trade-Offs Between Distance and Size of Towns

siderable variation in the prices of goods and services from one center to another results; and that regularly shaped market areas disappear.

Central Place Theory and the Location of Production

The principles underlying the central place theories of Christaller and Lösch were essentially economic ones. The businesspersons who provided the goods and services were assumed to be seeking to realize their largest possible profits while the farming population or consumers were assumed to be seeking to minimize their purchase costs by keeping their transportation costs as low as possible. Lösch, we noted, developed these economic principles in a more formal manner than did Christaller and explored the interrelations between demand levels, shapes of market areas, and network arrangements. But in his proposed system of superimposed networks in which different centers had varying combinations of economic functions, Lösch also provoked greater criticism from later economists. Recall that at the start of Lösch's discussion, the networks of market areas for particular economic functions were derived on the assumption that the pattern of demand was uniform over the region. But, as the critics have pointed out, this assumption was violated by the final location pattern of economic activities that resulted from Lösch's superimposition and rotation of the various networks. Lösch recognized this fact himself when he referred to "city-rich" and "city-poor" sectors.

The challenge for later economists has been to show that the systems of networks and cities derived by Christaller and Lösch are truly the best or optimal ones when measured by certain standards (for example, the minimizing of costs,) and when certain conditions hold in regard to the character of the region and its economy.

For the most part, these later contributions have been more mathematical in their contents than the earlier ones; not only does this make them more difficult to read and summarize, but it also imposes restrictions on the scope of the analyses themselves. In other words, the demonstration in formal mathematical terms that an arrangement or system is an optimal one becomes exceedingly difficult and complex if more than even one measure is employed. Consider the following example, provided by Tinbergen (1968).

Assume that there is a regional economy, which does not export or import any goods and which produces a number of products, h (where h runs from 0 up to some number H). One of these economic activites, (h=0), is agriculture, and this is carried on by an evenly distributed rural population. For each of the other economic activities there is a given size of firm or operation that is assumed to be the best economic size. In addition, each activity, it is assumed, has a given share of the total regional income, which is also set equal to total demand in the region. The number of firms or units, n_h, required to meet the demand for each activity can then be calculated and this, in turn, allows for a ranking of the economic activities with the lowest rank being given to the activity requiring the largest number of firms or plants ($n_1 \leq n_2 \leq n_3 \ldots \leq n_h$). This ranking establishes also the ordering of the urban centers. Tinbergen assumed that there would be as many types of urban centers as there were types of economic activities (H in his presentation) excluding agriculture, and at the top there would be an urban center in which the one enterprise or firm carrying on the highest-order economic activity ($n_h = 1$), would be located. Below this center, any other type of center of a particular rank, say rank 4 for example, would carry on all economic activities up to and including that rank, which would mean in this example activities 1, 2, 3, and 4. Further, Tinbergen assumed that in such a center of rank 4, for example, there would be just sufficient firms and output to satisfy the local demand in that center for the products of industries 1, 2, and 3. By contrast, in the case of industry 4, the center in question would have to serve not only the local demand but also that which existed in the lower-ranked centers. For a rank-4 center then, industry 4 would be an export industry. By the same token, in a center of rank 10, industry 10 would be its export activity. All centers at a particular rank would share equally in the export business for the products of the industry of that rank. The same argument would hold true for the industries and centers at any other rank.

With these particular features of a regional economy spelled out and for certain conditions having to do with the size of H and the nature of transportation costs, Tinbergen was able to demonstrate formally—that is, by way of mathematical proofs—that the hierarchical system of centers was an optimal or best possible one. Specifically, he proved that for the simple case when $H = 2$, the hierarchy model would result in the least possible total of the costs of transporting products within the region, assuming that either transport costs depended only on the type of good being transported and not on distance and so on, or that they depended on the dominant flow between any pair of centers and on the types of goods.

This approach toward demonstrating the optimality of a particular hierarchical arrangement of urban centers in which certain activities are located has been extended by others. Bos (1965) developed the analysis further in his book on *The Spatial Dispersion of Economic Activity*. His central place system was set within a region that had agriculture spread continuously throughout it, a region that required the production of other goods and services that were not perfectly divisible into small units of production and hence were subject to economies of scale, and a region within which transportation involved costs. On the last point, Bos considered three possibilities:

(1) Transportation costs depended only on the types of goods being shipped, and not on the distance over which the shipment was made;

(2) as a variation on (1), asuming the same transportation rates apply by types of goods, the cost of transportation between any pair of urban centers would then equal the higher cost of the trade flows between the centers;

(3) transportation costs depend not only on the types of goods being transported but also on the distances the goods are shipped.

For these different situations, Bos sought to determine whether the Tinbergen-type hierarchy of urban centers was an optimal one in terms of minimizing the total transportation cost. For the first case, it proved to be so. But for the second case no such confirmation was possible. Only for certain circumstances—notably where agriculture was carried on along with only two non agricultural activities and where both of these activities within a center were exporting activities (recall that in the Tinbergen model only the highest ranked one was an exporter)—were results obtainable. One such result was that if all three products (the agricultural and the two non agricultural ones) had positive transportation costs then the Tinbergen hierarchy of centers would never be an optimal one.

For the third situation concerning transportation costs, a number of conclusions followed depending on the relative levels of transportation costs for different industries. Bos concluded that the Tinbergen hierarchy would be an optimal one only if the lower-ranked industries had relatively high transportation costs.

This formal mathematical treatment of the question of how industries are optimally distributed among a hierarchy of urban centers can be extended even further, as Gunnarsson (1977) has demonstrated. Using a mathematical technique known as programming, he has shown how for a given number of urban centers and a specified number of industries a solution can be derived that not only ensures that the total cost of transporting goods among the centers is minimized but also determines the rankings of both the centers and the industrial sectors. Furthermore, if the assumption of fixed plant capacities for the industries is dropped, then he shows how it is possible to solve for not only the location of an industry but also its size. In this sense, the problem addressed by Gunnarsson is a version of that which is the subject of industrial location theory, the subject of another book in this series (Webber 1984).

In his discussion, Gunnarsson also bridges another gap, that between theory and planning. For in considering the optimal solution obtained from the hierarchical problem, the question arises as to whether or not the solution is one that can be interpreted as the result of the actions of individual decision makers who have information about prices. It is from this viewpoint of the individual entrepreneur that classical industrial location theory typically proceeds. Gunnarsson's conclusion, admittedly based on a discussion of certain of the mathematical properties of his model, was that "an optimal system of centers can only be maintained under decentralized decision-making when there is a central planning organ" (1977, p. 13). This central authority would be charged with the responsibility for either subsidizing certain forms of economic activity or taxing others in order to maintain and sustain that optimal allocation pattern of particular industries assigned to particular centers. Without such intervention, Gunnarsson concluded, undesirable tendencies such as the concentration of more activities in the larger centers than is consistent with the optimal solution (that is the minimization of total transportation costs in the system) may result. Of course, in the real world it is the case that multiple objectives are involved and optimization with respect to any single criterion such as transportation costs is seldom feasible. Nevertheless, Gunnarsson's work shows how an interest in theory can lead not only to challenging mathematical problems in regard to demonstrating the existence of optimal solutions, but can also throw useful light on a range of applied questions.

The Dynamics of Central Place Systems

The patterns of change over time in central place systems have been a subject of interest to many writers. Christaller himself spent a large section of his study on the discussion of the changes that would result from changing levels of income, improved transportation routes, and so on. I have referred already to the study by Parr (1980) in which structural change within a central place hierarchy was incorporated into a general central place model by way of a variable k value. There have been studies too, of which Morrill (1963) is a good example, in which simulations have been made of changing central place patterns. Morrill's concern was for the distribution of towns in Sweden as it had developed throughout the nineteenth and twentieth centuries. His simulation model began with the distribution of population in 1810, and then for particular time periods (generally 20-year intervals), the urban pattern was built up or simulated according to certain rules. Transport links were assigned first; then new manufacturing activities; then central place functions; and then, after the population distribution had been adjusted in an appropriate manner, the process was repeated. Each step involved the use of a probability model which assigned to each small subregion within the country a certain probability of receiving a transport link, a manufacturing activity, a central place activity, and a migrating population. The actual determination at each step of what was received and by which subregion was made by sampling numbers from a set of "random numbers." This procedure is known as the "Monte Carlo" method and we will not dwell on it further. Morrill concluded that his model results were reasonably realistic.

In line with the more mathematical treatment of central place theory noted in the preceding section, White (1977, 1978) has tackled the challenge of fashioning mathematical models of central place growth and decline. The approach adopted by White was to consider the size of a central place at a certain time as a function of its past size and its "profit ability." The latter concept was defined in terms of the difference between the revenue generated by all of the center's activities and the corresponding costs that they had to meet. The rationale was that the higher the profitability, the greater would be the center's growth.

It is, of course, one thing to write out an equation describing the growth of an economic sector in a central place (and in White's presentation there are as many such equations as there are economic sectors times the number of centers), but quite another thing to obtain any useful results from the exercise. To begin with, the number of such equations required to describe any actual central place system would be so large as to prohibit obtaining a mathematical solution. Also, the accounting problems to be faced in determining revenue and cost figures would make it an extremely costly exercise, even if it were possible to obtain solutions for a real-world situation.

The way out of this dilemma for White was to consider some of the formal properties of the model and, by using artificial data, to examine their significance in determining the model's outcomes. For example, in the model, "revenue" was defined in part as a function of the distance between a central place and the areas surrounding it in which the farm population or consumers resided. This distance factor, in turn, was weighted by a value that was meant to reflect the willingness to travel to shop for the particular good or service in question. (In the parlance of migration and shopping behavior studies this value is known as the "distance exponent" in the interaction equation). In simulations of the model's outcomes, different values were assigned to this distance weighting factor and their effects upon the population sizes of the centers were noted. It happened that these effects were quite pronounced, more so than those associated with, for example, the actual form of the interaction equations or the spatial arrangement of the centers.

Formal Restatements of the Central Place Theory

The extensions of the central place theory that we have discussed in the preceding two section share a common approach. They begin by defining a set of objects or things (towns, economic activities, farm households), they describe certain features or properties that these things possess (for example, size, plant capacity, incomes, budgets), and then they seek to derive or deduce certain additional properties of the objects (for example, optimal size distributions, least cost location patterns, and economic hierarchies). This is known as the *axiomatic method*. It is not basically different from the methods that Christaller and Lösch used except that the statements about both the objects and their properties are now very precise and detailed ones and the deductions are made using the tools of mathematics. Logic and formal proofs replace the maps and diagrams of the earlier studies. This formal axiomatic approach to the construction of a theory of central places found its strongest expression in two studies published by the Northwestern University Department of Geography (Dacey et al. 1974, Alao et al. 1977).

The first of the studies addressed the special case (the authors preferred to call it the "degenerate case") of a system of central places located along a line rather than on a plain. This was referred to as a one-dimensional central place system. The second study dealt with the more familiar Christaller-type central place system.

In both studies the approach was the same. The central place system was interpreted in three broad ways. It was first of all viewed as a geometric scheme in which there were points (the central places), regions (the trading areas), and networks (the arrangement of the points in relation to the regions). The mathematical properties of this scheme were examined. The geometrical system was then given an economic interpretation with economic activities and firms being assigned to the points (places), demands and prices being introduced, and the conditions of economic equilibrium being discussed. Finally, the central place system was examined from the point of view of its population characteristics and the patterns of migration and diffusion that took place within it.

In all three examinations, the language used was that of mathematics and the results were given in the form of theorems, lemmas, and proofs. Unfortunately, they are far too technical and numerous to reproduce here. Instead, a simple sketch is provided of the approach.

A set of central places can be defined as a set of points on a lattice or network. The hexagonal network is a particular form of this arrangement (Figure 5.2), and the properties of this network in terms of the distances between points and the sizes of the network cells can be specified. It is possible also to define sublattices in terms of the smaller sets of points that lie within them and to describe certain ordering principles by which these sublattices relate to one another.

This examination of the geometric structure does not require anything to be said about what functions the points or centers perform or what their relative sizes are. In this respect, the purely geometric structures described in these works on central place theory may not be unique and may be duplicated by similar structures in the physical world, for example in crystal structures. It is when the geometric concepts are interpreted in economic terms that the central place formulations of Christaller and Lösch come more clearly into focus. The argument goes as follows: Assume that income and hence demand are evenly distributed throughout the region and that there are firms with identical costs producing a single product for sale in the region. Can it be proven that there is a unique set of firm locations, outputs, and product prices that will result and will persist so long as nothing in the system changes? The answer is yes, and this solution defines an *economic equilibrium*. Now, can it be shown further that one equilibrium set of locations for the firms is the hexagonal network of points discussed earlier? Again, the answer is yes! For the firm in such a system, total transport cost, average cost of production, and average delivered price are minimized, and total quantity of goods supplied is maximized. For a market region in the same system, the total quantity of goods delivered and the overall satisfaction or *utility* are maximized.

Having proven the existence of an equilibrium solution for any one industry located over an hexagonal network, is it possible to show that together a set of such industries will form a Christaller-type central place system? In other words, is it possible to prove the existence of an equilibrium when industries are not acting independently of one another? The results in this case are generally positive ones except that it is not possible to prove that there is any one particular spatial arrangement of the economic activities over the hexagonal network that is the best possible one.

88

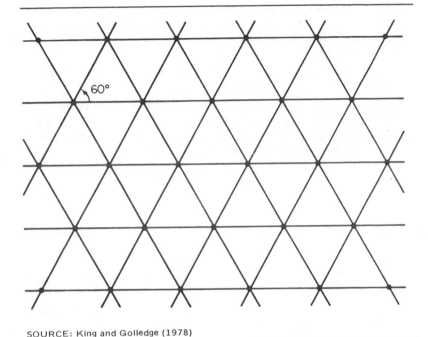

60°

SOURCE: King and Golledge (1978)

Figure 5.2 Hexagonal Network of Points

Up to this point, nothing has had to be assumed or derived concerning the population sizes of the centers in the system or concerning the levels of trade and other flows between them. The arguments have proceeded simply on the basis of statements about point locations, firms, market regions, and demands. Now if certain additional assumptions are made about the sizes of the rural and urban populations and about the relations between the levels of goods and services produced in the central places and the demands originating in the rural areas, then it is possible to derive results about the population sizes of the central places in the system. Furthermore, if the population is assumed to be migrating from rural regions to urban centers and, in turn, from urban centers to places of lower or higher orders, then formal results about the structure of migration within a Christaller system can be obtained.

These lines of formal mathematical reasoning, which we have only signposted here, often seem far removed from the reality of settlement patterns and urban activities. But they are, all the same, important illustrations of how rational human thought develops. Christaller and Lösch suggested that there were relationships between the number, size, functions, and locations of urban settlements. The empirical or observable support for their contentions is there for all of us to see and interpret. But even more fascinating is the intellectual challenge of examining the reasonableness of their assertions. Do they form a logically sound set of propositions and, when taken together, do they yield the set of conclusions or predictions that Christaller and Lösch suggested? Alao and his colleagues (1977) have shown that this is a most difficult question to answer.

Alternate Approaches to Settlement Theory

In their continued emphasis upon least-cost locations and profit-maximizing behavior on the part of firms, even the most mathematical treatments of central place theory have not departed radically from the general frameworks provided by the early works of Christaller and Lösch. The underlying premise is that there are general principles, perhaps even laws, of economic behavior that determine the character of the system and that can be stated precisely. These principles or laws are regarded as *deterministic* ones in the sense that their application to a given situation leads to a precisely determined outcome.

One feature of settlement systems in different parts of the world that everyone is agreed on is that they differ, and that even in those relatively homogeneous areas that approach the ideal conditions postulated by Christaller and Lösch, pure hexagonal and hierarchical systems of center do not occur.

Observations such as these have prompted some scholars to suggest that wholly different approaches are needed to the interpretation of settlement patterns. Two such options are mentioned here.

The first is represented in the many writings of Michael Dacey (1964, 1965, 1966a, 1966b). Dacey discarded all of the economic trappings of the central place theory and concentrated only on the location pattern of the settlements. This he regarded as the outcome of one or more *probabilistic processes*. In the case of such a process, for example, the tossing of a coin or the rolling of a dice, the outcome cannot be predicted precisely but can be anticipated only with a certain probability. There is, for example, a probability of 1/2 that a head will be the outcome on any toss of a normal coin. Dacey viewed the location pattern of settlements as the outcome of similar processes. One example will suffice, the so-called "county-seat" model (Dacey 1966a). Consider a large region divided up into a number of smaller subregions, the counties. Assume that by a random process a number of towns are distributed over these counties but in such a way that no one county can receive more than one town. It may receive none. Then, distribute over the same counties a second set of towns such that any county may receive none, one, two, or even all of this second set. Having made the two allocations, now consider the question, what is the probability that a county has a total of x towns? It turns out that this question can be answered and that the nature of the answer will depend upon the mathematical descriptions of the two processes that were used to distribute the two sets of towns over the counties. Again, we refrain from going into these mathematics. But the interesting point is that in his work, Dacey was able to derive predicted numbers of towns in answer to the above question that agreed quite closely with the observed number of towns per county in a state such as Iowa. The limitations of this approach, however, are not only that the economic issues are ignored completely but also that the towns are once again reduced simply to points and are not differentiated by size.

The second option to the classical central place theory that we shall mention here is that proposed by Leslie Curry (1964, 1967). Curry retained an economic framework, but like Dacey he introduced the operation of chance into the system. In Curry's model, both the consumer and the businesspersons were seen as acting in conditions of uncertainty and risk. In the case of the consumer, this meant that his or her demands for different goods and visits to different-sized centers both had an element of chance in them and could be described only as probability processes. In seeking to respond to this demand, the businessperson is faced with uncertainty about whether to stock and maintain inventories of certain goods. These different patterns of behavior and decision making are analyzed by Curry within a communication-theory framework, and from this he draws conclusions about the size of settlements, their economic bases, and their spacing.

6. CONCLUSION

The preceding chapters have sought to illustrate the richness of the legacy of Christaller and Lösch. Admittedly, in a world that is today so dominated by the patterns of economic, social, and political activity that are based in the large metropolitan centers of the world—New York, Tokyo, Shanghai, Moscow, London, Paris, and Berlin—it is difficult to give much credence to a theory that is so concerned with those smaller urban places that exist primarily to serve the inhabitants of rural regions. There are, however, several points that can be made in rebuttal of this argument.

First, the economic role of an urban center as one that provides goods and services to a surrounding region and to the smaller communities within it is one that is played by virtually all cities and towns. Every urban center is to some degree a central place and is subject to the forces and principles emphasized in the central place theory. These economic interrelationships come sharply into focus when, for example, political questions about metropolitan and regional government reorganization are raised. They surface also in many discussions of local taxation and financial issues.

Second, it has been shown that the concepts of hierarchies, nested trading areas, and so on have wider applicability than simply to rural areas. The organization and functioning of commercial districts within large metropolitan areas conform to many of the same principles. Further, it is not unreasonable to suppose that many of the same general trends and forces that have led to the decline of smaller hamlets and villages in the country-side—the increased accessibility and wider opportunities of the larger centers—might also be at work in regard to the smaller commercial center within the metropolitan areas.

Third, although it is true that rigid applications of the central place theory to problems in economic development planning have not been spectacularly successful, nevertheless the theory does provide several useful building blocks. In many Third World economies today, the overrapid urbanization is being concentrated in alarming proportions into one or two large cities, and it appears inevitable that catastrophes will result unless a more evenly balanced development of the countryside occurs. Whether or not this can be accomplished through a strengthening of the roles of smaller urban centers remains an open question. Central place theory is at least as relevant in this context as any other blueprint for regional development.

Finally, in the preceding chapter of this book we pointed to the intellectual challenges provided by central place theory and this is perhaps the best note to close on. The development of a theory is an intellectual exercise in which we seek to meld our observations of the real world (and they may be quite biassed observations when a social system is involved) with the rules of logic and reasoning. We seek to construct a logical framework, the theory, that not only gives meaning to those observations in hand but that will help us in understanding new situations and events. The intellectual challenge of fashioning such a framework, in which general rules about urban locational patterns, urban-size distributions, and economic activities are combined, is as intriguing today as it has ever been.

REFERENCES

Adams, J.W., and Kasakoff, A.B. 1976 Central-place theory and endogamy in China. In *Regional analysis, vol. 2, social systems,* ed. C.A. Smith, pp. 175-190. New York: Academic Press.

Alao, N. et al. 1977. *Christaller central place structures: an introductory statement.* Studies in Geography No. 22. Evanstown, IL: Northwestern University, Department of Geography.

Barnum, H.G. 1966. *Market centers and hinterlands in Baden-Wurttemberg.* Research Paper No. 103. Chicago: Univeristy of Chicago, Department of Geography.

Baskin, C.W. 1966. *Central places in Southern Germany.* Trans. of Christaller (1933). Englewood Cliffs, NJ: Prentice-Hall.

Beavon, K.S.O. 1977. *Central place theory: a reinterpretation.* New York: Longman.

Berry, B.J.L. 1963. *Commercial structure and commercial blight.* Research Paper No. 85. Chicago: University of Chicago, Department of Geography.

———. 1967. *Geography of market centers and retail distribution.* Englewood Cliffs, NJ: Prentice-Hall.

———. 1972. Hierarchical diffusion: the basis of developmental filtering and spread in a system of growth centers. In *Growth centers in regional economic development,* ed. N.M. Hansen, pp. 108-138. New York: The Free Press.

———. 1973. *The human consequences of urbanization.* New York: Macmillan.

Berry, B.J.L., Barnum, H.G., and Tennant, R.J. 1962. Retail location and consumer behavior. *Papers, Regional Science Association* 9: 65-106.

Berry, B.J.L., and Garrison, W.L. 1958. Functional bases of the central place hierarchy. *Economic Geography* 34: 145-154.

Berry, B.J.L., and Pred, A. 1961. *Central place studies: a bibliography of theory and applications.* Philadelphia: Regional Science Research Institute.

Borchert, J.R., and Adams, R.B. 1963. Trade centers and trade areas of the upper Midwest. Urban Report No. 3. *Upper Midwest Economic Study.* Minneapolis: University of Minnesota.

Bos, H.C. 1965. *Spatial dispersion of economic activity.* Rotterdam: University Press.

Braudel, Fernand. 1982. *The wheels of commerce.* London: W. Collins Sons & Co. Ltd.

Brutzkus, E. 1974. The scheme for spatial distribution of 5-million population in Israel. *The Developing Economies* 72: 47-49.

Buursink, J. 1981. On testing the nearest centre hypothesis. *Tijdschrift voor Econ. en Soc. Geografie* 72: 47-49.

Carol, H. 1960. The hierarchy of central functions within the city. *Annals of the Association of American Geographers* 50: 419-438.

Carroll, J.D. 1955. Spatial interaction and the urban metropolitan regional description. *Papers, Regional Science Association* 1: D1-D14.

Carter, H. 1972. *The study of urban geography.* London: Edward Arnold.

Carter, H., Stafford, H.A., and Gilbert, M.M. 1970. Functions of Welsh towns: implications for central place notions. *Economic Geography* 46: 25-38.

Christaller, W. 1933. *Die zentralen Orte in Süddeutschland.* Jena: Fischer.

Crissman, L.W. 1976. Spatial aspects of marriage patterns as influenced by marketing behavior in west central Taiwan. In *Regional Analysis: Vol. 2, social systems,* ed. C.A. Smith, pp. 123-148. New York: Academic Press.

Curry, L. 1964. The random spatial economy: an exploration in settlement theory. *Annals of the Association of American Geographers* 54: 138-146.

———. 1967. Central places in the random spatial economy. *Journal of Regional Science* 7: 217-238.

Dacey, M.F. 1964. Modified poisson probability law for point pattern more regular than random. *Annals of the Association of American Geographers* 54: 559-565.

———. 1965. Order distance in an inhomogenous random point pattern. *The Canadian Geographer* 9: 144-153.

———. 1966a. A county-seat model for the areal pattern of a urban system. *Geographical Review* 56: 527-542.

———. 1966b. A probability model for central place locations. *Annals of the Association of American Geographers* 56: 549-568.

Dacey, M.F., et al. 1974. One-dimensional central place theory. Studies in Geography No. 21. Evanston, IL: Northwestern University, Department of Geography.

Dickinson, R.E. 1947. *City, region and regionalism.* London: Routledge & Kegan Paul.

Forster, J.J.H., and Brummell, A. 1984. Multi-purpose trips and central place theory. *The Australian Geographer,* in press.

Galpin, C.J. 1915. The social anatomy of an agricultural community. University of Wisconsin Agricultural Experiment Station Research Bulletin No. 34.

Garner, B.J. 1966. The internal structure of retail nucleations. Studies in Geography no. 12. Evanston, IL: Northwestern University, Department of Geography.

Godlund, S. 1956. The function and growth of bus traffic within the sphere of urban influence. *Lund Studies in Geography, Series B,* No. 18.

Grove, D., and Huszar, L. 1964. *The towns of Ghana.* Accra: Ghana Universities Press.

Gunnarsson, J. 1977. *Production systems and hierarchies of centres.* Leiden: M. Nijhoff.

Hägerstrand, T. 1953. *Innovationsförloppet ur Korologish Synpunkt.* Lund: C.W.K. Gleerup. (Translated by A. Pred, 1967, as *Innovation Diffusion as a Spatial Process.* Chicago: University of Chicago Press.)

———. 1966. Aspects of the spatial structure of social communication and the diffusion of information. *Papers, Regional Science Association* 15: 27-42.

Harris, C.D. 1978. Patterns of cities. In *A man for all regions—the contributions of Edward L. Ullman to geography,* ed. J.D. Eyre, pp. 66-79. Chapel Hill: University of North Carolina, Department of Geography.

Harvey, M.D., Hung, M., and Brown, J.R. 1974. The application of p-median algorithm to the identification of nodal hierarchies and growth centers. *Economic Geography* 50: 187-202.

Hawley, A.H. ed. 1974. *Toward an understanding of metropolitan America.* San Francisco: Canfield Press.

Huang, J.C., and Gould, P. 1974. Diffusion in an urban hierarchy: the case of rotary clubs. *Economic Geography* 50: 333-340.

Hudson, J. 1972. Geographical diffusion theory. Studies in Geography No. 19. Evanston, IL: Northwestern University, Department of Geography.

94

Jefferson, M. 1931. Distribution of the world's city folks. *The Geographical Review* 21: 446-465.

Johnson, E.A.J. 1970. *The organization of space in developing countries.* Cambridge, MA: Harvard University Press.

Kant, E. 1951. Umland studies and sector analysis. *Lund Studies, in Geography, Series B.* 3: 3-13.

King, L.J. 1962. The functional role of small towns in the Canterbury area. *Proceedings, Third New Zealand Geography Conference: 139-149.*

King, L.J., and Golledge, R.G. 1978. *Cities space and behavior. The elements of urban geography.* Englewood Cliffs, NJ: Prentice-Hall.

Kolb, J.H. 1923. Service relations of town and country. *University of Wisconsin Agricultural Experiment Station Bulletin,* No. 58.

Kolb, J.H., and Brunner, E. de S. 1940. *A study of rural society.* 2nd. ed. Boston: Houghton Mifflin.

Lösch, A. 1940. *Die räumliche Ordung der Wirtschaft.* Jena: Fischer. Translated by W.H. Woglom and W.F. Stolper (1954), *The Economics of Location.* New Haven, CT: Yale University Press.

Mabogunje, A.L. 1980. *The development process.* London: Hutchinson & Co.

Marshall, J.U. 1969. The location of service towns. an approach to the analysis of central place systems. Research Publications. Toronto: University of Toronto, Department of Geography.

Morrill, R.L. 1963. The development of spatial distributions of towns in Sweden: an historical – predictive approach. *Annals of the Association of American Geographers* 53: 1-14.

Müller-Wille, C.F. 1978. The forgotten heritage: Christaller's antecedents. In *Perspectives in geography 3: the nature of change in geographical ideas,* ed. B.J.L. Berry, pp. 37-64. Dekalb: Northern Illinois University Press.

Murdie, R.A. 1965. Cultural differnces in consumer travel. *Economic Geography* 41: 211-233.

Parr, J.B. 1980. Frequency distributions of central places in Southern Germany: a further analysis. *Economic Geography* 56: 141-154.

Parr, J.B., and Denike, K.G. 1970. Theoretical problems in central place analysis. *Economic Geography* 46: 568-586.

Pred, A.R. 1973. Urbanization, domestic planning problems and Swedish geographic research. *Progress in Geography* 5: 1-76.

———. 1977. *City-systems in advanced economics.* London: Hutchinson and Co.

Preston, R.E. 1983. The dynamic component of Christaller's central place theory and the theme of change in his research. *The Canadian Geographer* 27: 4-16.

Proudfoot, M.J. 1937. The outlying business centers of Chicago. *Journal of Land and Public Utility Economics* 13: 57-70.

Pyle, G.F. 1969. The diffusion of cholera in the United States in the nineteenth century. *Geographical Analysis* 1: 59-75.

Rushton, G. 1971. Postulates of central place theory and the postulates of central place systems. *Geographical Analysis* 3: 140-156.

Robic, M.C. 1982. Cent ans avant Christaller. Une theorie des lieux centraux. *L'espace geographique* XI: 5-12.

Saey, P. 1973. Three fallacies in the literature on central place theory. *Tijdschrift voor Econ. en Soc. Geografie* 64: 181-194.

Saey, P., and Lietaer M. 1980. Consumer profiles and central place theory. *Tijdschrift voor Econ. en Soc. Geografie* 71: 180-186.

———. 1981. On testing Christaller's theory: a rejoinder. *Tijdschrift voor Econ. en Soc. Geografie* 72: 50-51.

Saskatchewan Royal Commission on Agriculture and Rural Life. 1957. Service Centers Report, No. 12.

Schachar, A.S. 1971. Israel's development terms: evaluation of a national urbanization policy. *Journal of the American Institute of Planners* 37: 362-372.

Simmons, J. W. 1964. The changing pattern of retail location. Research Paper No. 92. Chicago: University of Chicago, Department of Geography.

———. 1975. Canada: choices in a national urban strategy. Research Paper No. 70. Toronto: University of Toronto, Center for Urban and Community Studies.

Skinner, G.W. 1964. Marketing and social structure in rural China, part I. *Journal of Asian Studies* 24: 32-43.

———. 1965. Marketing and social structure in rural China, part II. *Journal of Asian Studies* 24: 195-228.

———. 1976a. Mobility strategies in late imperial China: a regional systems analysis. In *Regional analysis, vol. 1, economic systems*, ed. C.A. Smith, pp. 327-364. New York: Academic Press.

———. 1976b. Cities and the hierarchy of local systems. In *The City in Late Imperial China*, ed. G.W. Skinner, pp. 275-351. Palo Alto, CA: Stanford University Press.

Smith, C. 1976a. Regional economic systems: linking geographical models and socioeconomic problems. In *Regional analysis, vol. 1, economic systems*, ed. C.A. Smith, pp. 3-63. New York: Academic Press.

———. 1976b. Causes and consequences of central-place types in western Guatemala. In *Regional analysis, vol. 1, economic systems*, ed. C.A. Smith, pp. 225-300. New York: Academic Press.

Stafford, H.A. 1963. The functional bases of small towns. *Economic Geography* 39: 165-175.

Taaffe, E.J., et al. 1970. *Geography*. Englewood Cliffs, NJ: Prentice-Hall.

Thijsse, J.P. 1962. A rural pattern for the future in the Netherlands. *Papers, Regional Science Association* 10: 133-141.

Tinbergen, J. 1968. The hierarchy model of the size distribution of centers. *Papers, Regional Science Association* 20: 65-68.

United Nations. 1974. *Concise report on the world population situation in 1970-75 and its long-range implications*. New York: United Nations.

United States Bureau of the Census. 1980. *Census of populations: 1960, 1970, and 1980*, vol. 1. Washington, DC: U.S. Government Printing Office.

Vining, R. 1955. A description of certain spatial aspects of an economic system. *Economic Development and Cultural Change* 3: 147-195.

Webber, M.J. 1971. Empirical verifiability of classical central place theory. *Geographical Analysis* 3: 15-28.

———. 1984. *Industrial location theory*. Beverly Hills: Sage.

Weisbrod, R. 1974. Diffusion of relative wage inflation in Southeast Pennsylvania. Studies in Geography No. 23. Evanston, IL: Northwestern University, Department of Geography.

White, R.W. 1977. Dynamic central place theory: results of a simulation approach. *Geographical Analysis* 9: 226-243.

———. 1978. The simulation of central-place dynamics: two-sector systems and the rank-size distribution. *Geographical Analysis* 10: 201-208.

Woldenberg, M.J. 1968. Energy flow and spatial order: mixed hexagonal hierarchies of central places. *Geographical Review* 58: 552-574.

Wrigley, E.A. 1978. Parasite or stimulus: the town in a pre-industrial economy. In *Towns in societies*, eds. P. Abrams and E.A. Wrigley, pp. 295-309. Cambridge: Cambridge University Press.

ABOUT THE AUTHOR

LESLIE J. KING is a New Zealander by birth and a geographer by training. After completing his B.A. and M.A. degrees at the University of New Zealand in Christchurch, he travelled to the United States in 1957 on a Fulbright Award and completed a three-year Ph.D. program at the University of Iowa. He has subsequently held faculty appointments at his alma mater in New Zealand, at McGill University in Montreal, Canada, at the Ohio State University and, since 1970, at McMaster University in Hamilton, Ontario, Canada, where he has served as Chairman of the Geography Department, Dean of Graduate Studies and now, Vice President (Academic). Dr. King has published widely in the literature of urban and economic geography and regional science. He was the founding editor of *Geographical Analysis: An International Journal of Theoretical Geography* published by The Ohio State University Press. In 1976 he was awarded the Distinguished Service Award of the Association of American Geographers. His teaching and research interests are urban and economic geography, regional economics, regional science, and methodologies of the social sciences.